Bacterial Genetics and Genomics

Bacterial Genetics and Genomics

Lori A.S. Snyder

CRC Press
Taylor & Francis Group

A GARLAND SCIENCE BOOK

CRC Press
Taylor & Francis Group
6000 Broken Sound Parkway NW, Suite 300
Boca Raton, FL 33487-2742

© 2020 by Taylor & Francis Group, LLC
CRC Press is an imprint of Taylor & Francis Group, an Informa business

No claim to original U.S. Government works

Printed on acid-free paper

International Standard Book Number-13: 978-0-367-26376-8 (Hardback)
978-0-8153-4569-5 (Paperback)

Visit this book's Companion Web site:
www.crcpress.com/cw/snyder

Visit the Taylor & Francis Web site at
http://www.taylorandfrancis.com

and the CRC Press Web site at
http://www.crcpress.com

Contents in Brief

Contents

Chapter 3
Genomes   45

Chapter 8
Protein Folding and Structure   129

Chapter 9
Multiprotein Systems and Proteomes   143

Part V   Bacterial Response, Adaptation, and Evolution

Chapter 13
Bacterial Response   211

Chapter 14
Bacterial Adaptation   227

Chapter 15
Bacterial Evolution   241

Part VI   Gene Analysis, Genome Analysis, and Laboratory Techniques

Chapter 16
Gene Analysis Techniques   255

Part VII   Applications of Bacterial Genetics and Genomics

Chapter 19
Biotechnology   321

Note to Reader

This book is organized into chapters that are grouped into parts, made up of three chapters that go together. This organization is designed to help group key ideas together and to build on concepts presented in earlier chapters as more advanced topics are discussed in later chapters.

Whenever a new term is encountered, this will be shown in bold. The definition of the term will be in the text in context where it is used, but it will also be in the Glossary at the back of the book. The term will only be in bold the first time, so if there is an unfamiliar term, it is worth checking the Glossary for a definition.

In addition to a standard Glossary, there is also a Glossary of Bacterial Species at the back of this book. This has been included as a quick reference guide so that readers can easily find out about the bacterial species that are mentioned. There are 97 species referred to in this book and it is not expected that readers will have heard of all of these. Bacterial genetics and genomics has developed far beyond *Escherichia coli*, which is reflected in the writing here.

Where possible, interesting and useful information is also explained in figures and sometimes in tables or other illustrations. These will be accompanied by explanatory text that is slightly different from the main text. By explaining the concepts in two slightly different ways and including an illustration, this should clarify these points.

Each chapter ends with a list of the Key Points covered in the chapter. This list can be used to determine if the topics have all been clearly understood, before moving on. It also summarizes what has been covered in the chapter.

At the end of each chapter there are also some questions about the topics and terms covered in the chapter to aid in study and revision. There are three types of questions. First, terms introduced in the chapter are reviewed to ensure understanding. Next, questions are asked about the material in the chapter to test understanding of what has been read. Finally, some topics are presented for discussion, which extend the subject beyond what has been covered in the chapter itself.

Lastly, each chapter ends with a list of Further Reading. These are provided for students who would like to do some further reading in the subject area. The references provided may be useful in completing the discussion questions at the end of the chapter and in satisfying curiosity.

As a useful reference and guide, other information that might come in handy when studying and applying your knowledge of bacterial genetics and genomics is printed on the inside of the covers of this book.

Companion website

The companion website to this book, www.crcpress.com/cw/snyder, contains the following:

- Links to external sites which support Chapters 16 and 17.

- Answers to the end-of-chapter questions.

- Hints to the discussion questions.

- Flashcards from the online glossary.

- Figures from the book available for download (as PDF and PPT).

- Link to affiliated blog, which provides information on advances in bacterial genetics and genomics.

Acknowledgments

Throughout the writing of this book I have received valuable assistance from my editors, Liz Owen and Jordan Wearing, who have diligently read through my chapters, providing feedback, formatting advice, and guidance. Thank you. Thanks as well to the other staff at Taylor & Francis who made this publication possible, including Chuck Crumly, and the illustrator, Patrick Lane, who has brought my pencil sketches to life. I would also like to thank the reviewers— those who provided helpful comments on the original proposal for *Bacterial Genetics and Genomics* and for those who provided detailed feedback on the written chapters.

Finally, special thanks go to my husband Alex and my daughter Wilhelmina for their patience with me as I did most of my writing at home from what became referred to as "my nest." Your support and love are very much appreciated.

About the Author

I grew up in Clinton, Connecticut, USA, and it is there that my interest in biology started, due in large part to my parents. My mother and father met at Yale University Hospital in New Haven, Connecticut, where he was a respiratory therapist and she was a nurse. Our conversations over family dinner made it clear to me that the science lessons I was learning in school had direct application in the real world. In middle school, my father and I made a spirometer for measuring the vital capacity of the lungs in our basement workshop. When it came time for me to decide on universities, my mother and I took a road trip to visit the top places on my list, which solidified my decision to go to the College of William and Mary in Virginia, USA. There I studied both Biology and Music. When I started, I was determined to be a veterinarian, and also to continue to study guitar, which I had been playing since I was 10 years old. As I started learning more in my course, I developed a passion for genetics and decided to do a PhD. It was recommended that I first do a Master's degree, due to the competitive nature of human genetics PhDs. So, I decided to stay at William and Mary for a two-year program, which came with a stipend and paid my tuition fees. A condition of the stipend was that I taught laboratory classes; this was an excellent opportunity to gain some teaching experience. I wasn't convinced I would enjoy teaching before I started, but I found it very rewarding and ever since then I have sought to find a balance in my career between teaching and research, both aspects of academic life that I enjoy. It was during my Master's degree that I started my research in bacteriology in a class with Professor Carl Vermeulen. The summer before my class with Dr. V, one of his students had made a discovery and he offered us the chance to pursue this discovery as a research project. Because of this opportunity, I embarked on my microbiology career, working on *Escherichia coli* lipopolysaccharide. I was able to present this research at the American Society for Microbiology General Meeting, an awe inspiring experience due to the size and scope of the conference. When I finished my Masters, I applied for PhDs that had strong microbiology and molecular genetics aspects. I found this at Emory University in Atlanta, Georgia, USA. As before, this came with a stipend and my fees paid, in exchange for a bit of teaching. I found an ideal home in which to do my PhD with my supervisor, Professor Bill Shafer. His group was working on antibiotic resistance, which was a subject important to me, and his research involved molecular genetics. The subject of the research was perfect. My short rotation in the first year went really well. I got some good results and got along well with the other people in the lab and with Bill. My decision to complete my PhD in his group has shaped my career going forward. His research focused on the bacterial species *Neisseria gonorrhoeae*, which has remained my focus. A year or so after I started my PhD, the first few fragments of the genome sequence data of this species arrived in Bill's lab and I started doing the analysis. This began my love of bacterial genetics and genomics, which has been a common thread in my research ever since, ultimately leading to the writing of this book. As I was writing my PhD thesis for Bill, I remember him remarking on the quality of my writing and that I might have a future in academic writing. I have always carried that supportive comment from him with me and it has been an inspiration throughout the writing of this book and other works. Since my PhD at Emory, I have been involved in several genomics projects at the University of Oxford, UK, and University of Birmingham, UK, during my post-doctoral training and now at Kingston University, where I have been a Senior Lecturer, Reader, and Associate Professor.

Lori A.S. Snyder
April 2020

Welcome to the World of Bacterial Genetics and Genomics

Bacteria are small, single-celled organisms that play various important roles on the planet and in the lives of human beings. Although they are microscopic, bacteria have had a tremendous influence upon the other organisms of the world, which are often more complex in their cellular make up and in their genetics.

The study of genetics is the study of the code of life that is carried within the DNA of living things. Bacteria have been the subject of extensive research into their genetics, both because of the roles for some of these in human disease and also because these organisms are easy to grow, easy to manipulate, and therefore a great deal can be learned about the principles of genetics.

Our understanding of bacterial genetics and the ways in which we investigate bacterial genetics has changed as the genomics field has advanced. Genomics is wider in scope than genetics, being concerned with all of the DNA in an organism, rather than a single gene or small portion. These two research fields, bacterial genetics and bacterial genomics, complement each other and influence each other. Where once we did not know that DNA was the genetic material, now we can lay bare the whole genome of any bacteria using sequencing methods. Genetics and genomes can no longer be separated, and under the novel insights from genetics and genomics, once-believed borders in biology start to fade: most biological knowledge can be viewed under a new light when applied to the bacterial world.

Once it was believed that genome size correlated with the complexity of the organism. Now we know that some viral genomes are not only larger than some bacterial genomes, but also more complex. Some genomes of higher organisms (eukaryotes) are smaller than some bacterial genomes, challenging our old assumptions. Bacterial plasmids can be as big as chromosomes and can carry essential genes, required by the bacteria, just like chromosomes. The difference between a bacterium and an organelle within a eukaryotic cell are blurred as further evidence supports organelle evolution from bacteria. Genomics has even revealed the transfer of genes between bacteria and higher organisms. Bacteria can even be larger than eukaryotic cells and can form multicellular life forms, once believed to be the realm of higher organisms. There are exceptions to the standard rules that are thought of as differentiating bacteria and eukaryotes, including the identification of bacteria that contain linear chromosomes.

The species concept, generally relying on sexual reproduction to classify different species, cannot accurately describe the differences and relations within and between groups of bacteria. Bacterial cells do not undergo sexual reproduction and their genomes can be influenced by horizontal gene transfer from unrelated bacteria. Although we define bacteria into species, largely identified by characteristics observed in the laboratory, in the field, or outcomes of human infection, the whole concept of what constitutes a species is hotly debated when dealing with bacteria.

The bacterial world is incredibly diverse. Assumptions made about the lines between bacteria, viruses, and eukaryotes have been challenged. Similarly, assumptions that all bacteria behave the same must be cast aside. Even within a bacterial species, there can be significant variation in the genes within the genome, the expression pattern of those genes, and their behavior. These concepts will be introduced and explored, both through comparing different bacterial species and exploring the diversity that can be seen within species and within populations.

Further reading

Ku C, Nelson-Sathi S, Roettger M, Sousa FL, Lockhart PJ, Bryant D, Hazkani-Covo E, McInerney JO, Landan G, Martin WF. Endosymbiotic origin and differential loss of eukaryotic genes. *Nature.* 2015; 524(7566): 427–32.

Raoult D, La Scola B, Birtles R. The discovery and characterization of mimivirus, the largest known virus and putative pneumonia agent. *Clin Infect Dis.* 2007; 45(1): 95–102.

Vivarès CP, Méténier G. Towards the minimal eukaryotic parasitic genome. *Curr Opin Microbiol.* 2000; 3(5): 463–7.

Part I

DNA, Genes, and Genomes

DNA

This chapter will introduce the biologically important molecule, **DNA**, **deoxyribonucleic acid**. DNA is the genetic material, carrying within its structure the code of life, thus it is important to understand how this molecule arose on Earth, how scientists determined that DNA held the code of life, what the structure of the molecule is, and how this holds the genetic information. Terms and concepts that will be explored in more detail in this book are introduced here, forming the fundamental background and concepts of bacterial genetics and genomics.

Life originated from RNA with DNA evolving later

There once was a time when there was no DNA (deoxyribonucleic acid). Billions of years in the past, before life emerged on the planet, a primordial soup of chemicals and compounds held the promise of what we have on the Earth today. In our study of genetics and genomics today, we analyze the sequence of DNA, but when investigating the distant past, it is believed that it was not DNA that was at the origin, but **RNA** (**ribonucleic acid**).

According to the **RNA World hypothesis**, the first nucleic acids on our planet were RNAs, which were first established as catalyst and genetic material. Early life on Earth is believed to be RNA-based, rather than DNA. Today, there exist viruses with RNA genomes that serve as examples and evidence that RNA can be the carrier of the genetic code. The discovery of **ribozymes**, RNA enzymes (covered in Chapter 4), has provided further supporting evidence for the RNA World hypothesis by showing that RNA can be a biological catalyst. Therefore, the necessary functions for early life on Earth can all have been accomplished by RNA in the absence of DNA. DNA-based storage of genetic information, which is seen in living cells today, came later in the evolution of life. The RNA World model, with DNA arising on Earth later, is supported by chemistry and the complexity of enzymes needed for DNA to be synthesized and therefore used as the code of life.

It is believed that DNA arose after templated protein synthesis, whereby proteins are made based on the sequence of RNA. There are some suggestions that DNA may have emerged before this process arose, although these theories still start with RNA. It is also unclear whether DNA came about before or after the **Last Universal Common Ancestor** (**LUCA**), that is, the lifeform that came before the lineage split that resulted in today's **bacteria**, **eukaryotes**, and **archaea**. If DNA arose after the Last Universal Common Ancestor, this would explain fundamental key differences in the way in which errors are corrected in the genetic code between the different lineages. If, however, DNA came about before the bacteria, eukaryotes, and archaea lineages split, then the differences between the genes responsible for accurately copying DNA in bacteria versus the other lineages would have to be due to the complete replacement of these genes in the other lineages later in evolution. If DNA containing these genes and other features was present before the split, then it would follow that these genes should continue to carry similarity with one another across the three lineages. Because they do not, either DNA arose after the Last Universal Common Ancestor or it arose before and the copying accuracy genes were later swapped out for those with similarity to what is seen in the lineages today.

Over 4 billion years ago, the first life on Earth appeared. This first life form was the initial replicator, capable of making more of itself, and it was the beginning

Figure 1.1 Evolution of life as we know it according to the RNA World hypothesis. The world we understand now evolved from an earlier form. One hypothesis, strongly supported, is that life began as an RNA World. It is believed that 4 billion years ago, the Initial Darwinian Ancestor arose, the first life form that started evolution on Earth. From this later arose the Last Universal Common Ancestor, which diverged into the lineages we describe today as bacteria, archaea, and eukaryotes. There was a transition to the world of today, a DNA/RNA/protein world, either before or after the Last Universal Common Ancestor.

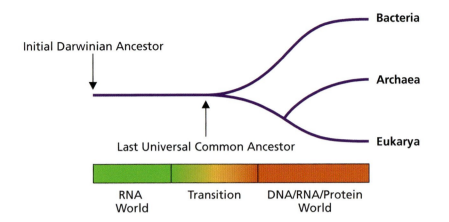

of evolution as the **Initial Darwinian Ancestor** (**Figure 1.1**). The Last Universal Common Ancestor came much later. The RNA World hypothesis would mean that Darwinian Evolution occurred in two phases: first as the RNA World of ribozymes catalyzing metabolism and RNA replication starting with the Initial Darwinian Ancestor; and then as the DNA/RNA/protein world of DNA, rRNA, mRNA, tRNA, ribosomes, and protein enzymes for metabolism, RNA synthesis, protein synthesis, and DNA replication, which continues today. Evolution, based on RNA, would have occurred before DNA arose and before the split into the prokaryote, eukaryote, and archaea lineages, in the first form(s) of life on Earth. As these progressed, evolution led to the Last Universal Common Ancestor.

Nucleic acids are made of nucleoside bases attached to a phosphate sugar backbone

The complexity of life is determined by **nucleic acids**. These biological molecules are what make up DNA and RNA, the blueprints for life. All nucleic acids have two aspects: the **backbone** and the **nucleoside bases**. The nucleic acid backbone itself is made of two parts, alternating between five-carbon sugars and phosphates that form a chain. Each sugar in the backbone chain is attached to a nucleoside base, also referred to as a **nitrogenous base** due to the inclusion of nitrogens in the structure.

There are four different nucleoside bases that can be included in each nucleic acid molecule, which in turn can have hundreds, thousands, or millions of bases depending on its length. The order, pattern, and combination of the nucleoside bases determine the genetic information that is carried by the nucleic acid.

RNA contains ribose sugars and is single-stranded

The first nucleic acid on the planet, according to the RNA World hypothesis, was RNA. The sugar within the RNA backbone is **ribose**, a monosaccharide that is present here in a five-member ring form (**Figure 1.2**). The nucleotide bases attached to the ribose are **adenine** (A), **cytosine** (C), **guanine** (G), and **uracil** (U). These bases are either **purines** (A and G), which have two rings in their structure, or **pyrimidines** (U and C), which have one ring. RNA is single-stranded, meaning that it has one backbone, although in some cases the RNA molecule may bend back upon itself, forming structures that are necessary for the function of some RNAs. RNA is the genetic material of some viruses and **bacteriophages**, viruses that infect bacteria, encoding the whole of their genome without the need for DNA. Even in cells with DNA genomes, RNA is important for protein synthesis. It is the molecule that is the intermediate between DNA and protein in the biosynthetic process of expressing the code of life in DNA as functional and structural components of the cell in proteins. RNA is the template by which the order of amino acids in a protein is directly interpreted.

DNA

This chapter will introduce the biologically important molecule, **DNA**, **deoxyribonucleic acid**. DNA is the genetic material, carrying within its structure the code of life, thus it is important to understand how this molecule arose on Earth, how scientists determined that DNA held the code of life, what the structure of the molecule is, and how this holds the genetic information. Terms and concepts that will be explored in more detail in this book are introduced here, forming the fundamental background and concepts of bacterial genetics and genomics.

Life originated from RNA with DNA evolving later

There once was a time when there was no DNA (deoxyribonucleic acid). Billions of years in the past, before life emerged on the planet, a primordial soup of chemicals and compounds held the promise of what we have on the Earth today. In our study of genetics and genomics today, we analyze the sequence of DNA, but when investigating the distant past, it is believed that it was not DNA that was at the origin, but **RNA** (**ribonucleic acid**).

According to the **RNA World hypothesis**, the first nucleic acids on our planet were RNAs, which were first established as catalyst and genetic material. Early life on Earth is believed to be RNA-based, rather than DNA. Today, there exist viruses with RNA genomes that serve as examples and evidence that RNA can be the carrier of the genetic code. The discovery of **ribozymes**, RNA enzymes (covered in Chapter 4), has provided further supporting evidence for the RNA World hypothesis by showing that RNA can be a biological catalyst. Therefore, the necessary functions for early life on Earth can all have been accomplished by RNA in the absence of DNA. DNA-based storage of genetic information, which is seen in living cells today, came later in the evolution of life. The RNA World model, with DNA arising on Earth later, is supported by chemistry and the complexity of enzymes needed for DNA to be synthesized and therefore used as the code of life.

It is believed that DNA arose after templated protein synthesis, whereby proteins are made based on the sequence of RNA. There are some suggestions that DNA may have emerged before this process arose, although these theories still start with RNA. It is also unclear whether DNA came about before or after the **Last Universal Common Ancestor** (**LUCA**), that is, the lifeform that came before the lineage split that resulted in today's **bacteria**, **eukaryotes**, and **archaea**. If DNA arose after the Last Universal Common Ancestor, this would explain fundamental key differences in the way in which errors are corrected in the genetic code between the different lineages. If, however, DNA came about before the bacteria, eukaryotes, and archaea lineages split, then the differences between the genes responsible for accurately copying DNA in bacteria versus the other lineages would have to be due to the complete replacement of these genes in the other lineages later in evolution. If DNA containing these genes and other features was present before the split, then it would follow that these genes should continue to carry similarity with one another across the three lineages. Because they do not, either DNA arose after the Last Universal Common Ancestor or it arose before and the copying accuracy genes were later swapped out for those with similarity to what is seen in the lineages today.

Over 4 billion years ago, the first life on Earth appeared. This first life form was the initial replicator, capable of making more of itself, and it was the beginning

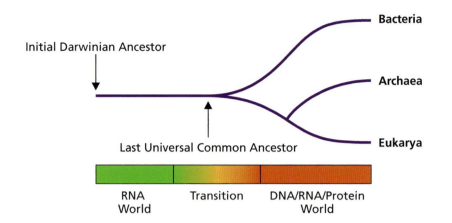

of evolution as the **Initial Darwinian Ancestor** (**Figure 1.1**). The Last Universal Common Ancestor came much later. The RNA World hypothesis would mean that Darwinian Evolution occurred in two phases: first as the RNA World of ribozymes catalyzing metabolism and RNA replication starting with the Initial Darwinian Ancestor; and then as the DNA/RNA/protein world of DNA, rRNA, mRNA, tRNA, ribosomes, and protein enzymes for metabolism, RNA synthesis, protein synthesis, and DNA replication, which continues today. Evolution, based on RNA, would have occurred before DNA arose and before the split into the prokaryote, eukaryote, and archaea lineages, in the first form(s) of life on Earth. As these progressed, evolution led to the Last Universal Common Ancestor.

Nucleic acids are made of nucleoside bases attached to a phosphate sugar backbone

The complexity of life is determined by **nucleic acids**. These biological molecules are what make up DNA and RNA, the blueprints for life. All nucleic acids have two aspects: the **backbone** and the **nucleoside bases**. The nucleic acid backbone itself is made of two parts, alternating between five-carbon sugars and phosphates that form a chain. Each sugar in the backbone chain is attached to a nucleoside base, also referred to as a **nitrogenous base** due to the inclusion of nitrogens in the structure.

There are four different nucleoside bases that can be included in each nucleic acid molecule, which in turn can have hundreds, thousands, or millions of bases depending on its length. The order, pattern, and combination of the nucleoside bases determine the genetic information that is carried by the nucleic acid.

RNA contains ribose sugars and is single-stranded

The first nucleic acid on the planet, according to the RNA World hypothesis, was RNA. The sugar within the RNA backbone is **ribose**, a monosaccharide that is present here in a five-member ring form (**Figure 1.2**). The nucleotide bases attached to the ribose are **adenine** (A), **cytosine** (C), **guanine** (G), and **uracil** (U). These bases are either **purines** (A and G), which have two rings in their structure, or **pyrimidines** (U and C), which have one ring. RNA is single-stranded, meaning that it has one backbone, although in some cases the RNA molecule may bend back upon itself, forming structures that are necessary for the function of some RNAs. RNA is the genetic material of some viruses and **bacteriophages**, viruses that infect bacteria, encoding the whole of their genome without the need for DNA. Even in cells with DNA genomes, RNA is important for protein synthesis. It is the molecule that is the intermediate between DNA and protein in the biosynthetic process of expressing the code of life in DNA as functional and structural components of the cell in proteins. RNA is the template by which the order of amino acids in a protein is directly interpreted.

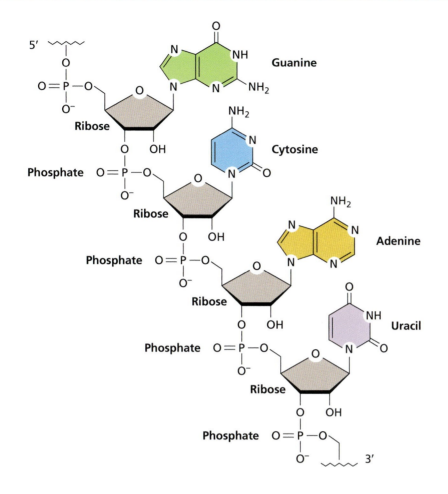

Figure 1.2 The chemical structure of RNA. The phosphates are joined to the ribose sugars by phosphodiester bonds to make up the backbone of the RNA. The nucleotide bases guanine, cytosine, adenine, and uracil, attach to the ribose sugar at carbon 1.

DNA contains deoxyribose sugars and is double-stranded

DNA stores the genetic information for everything, aside from some viruses and bacteriophages. In DNA, the sugar is **deoxyribose**, another monosaccharide with a five-member ring that is similar to ribose, although lacking one OH group. The nucleotide bases in DNA are **adenine** (A), **cytosine** (C), **guanine** (G), and **thymine** (T). In DNA, the pyrimidine thymine is present instead of uracil. These differences, the presence of the deoxyribose sugar and of thymine are characteristic of DNA and set it apart from RNA. DNA is double-stranded, meaning that there are two backbones going in opposite directions in a structure referred to as **antiparallel**. The directionality of each strand is defined by the bonds in the backbone on the 3rd (3′) and 5th (5′) carbons on the sugars. The direction of each strand impacts upon the processes of **replication**, the copying of DNA, and **transcription**, the generation of RNA based on the DNA sequence. The two strands of DNA are another key characteristic of this molecule. Although single-stranded DNA can be found in nature and generated in the laboratory, DNA's native form is double-stranded. In this form, the nucleotide bases covalently attached to one backbone are joined by **hydrogen bonds** to the nucleotide bases covalently attached to the other backbone (**Figure 1.3**). These hydrogen bonds between the nucleotide bases form the double-stranded structure. Adenine forms two hydrogen bonds with thymine and guanine forms three hydrogen bonds with cytosine on the opposite strand of the DNA structure. The double-stranded structure of DNA twists into a double helix (**Figure 1.4**).

The double helix of DNA is by now famous, included today in works of art as well as scientific textbooks. It has been the inspiration for sculptures (**Figure 1.5**), paintings, fashion, jewelry, tattoos, and other creative works. The DNA double helix has even been featured on the back of the UK £2 coin (**Figure 1.6**). The structural

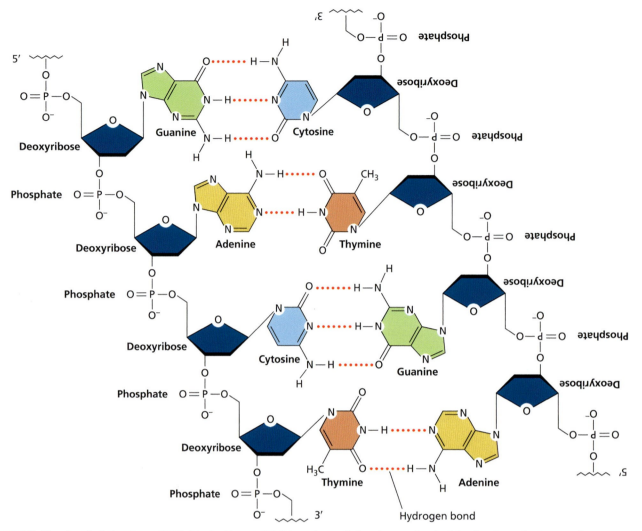

Figure 1.3 The chemical structure of DNA. The backbone of DNA is made up of phosphate molecules joined to deoxyribose sugars by phosphodiester bonds. Note that the deoxyribose sugars lack an OH group on carbon 2 that is present on the ribose sugars of RNA. The nucleotide bases guanine, cytosine, adenine, and thymine, attach to the deoxyribose sugar at carbon 1. Hydrogen bonds between complementary bases connect the two strands of the DNA structure.

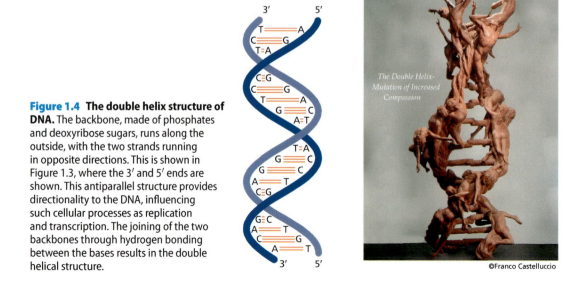

Figure 1.4 The double helix structure of DNA. The backbone, made of phosphates and deoxyribose sugars, runs along the outside, with the two strands running in opposite directions. This is shown in Figure 1.3, where the 3' and 5' ends are shown. This antiparallel structure provides directionality to the DNA, influencing such cellular processes as replication and transcription. The joining of the two backbones through hydrogen bonding between the bases results in the double helical structure.

Figure 1.5 The DNA double helix as an inspiration for art. This is "The Double Helix—Mutation of Increased Compassion" by Franco Castelluccio. The structure of the double helix is here inspiration for a sculpture balancing male and female forms as the backbone of DNA. Artistic works such as these inspire discussions about evolution, mutation, and metamorphosis. (Image courtesy of Franco Castelluccio.)

©Franco Castelluccio

model of the DNA double helix created by James Watson and Francis Crick when they first unraveled the structure of the molecule from its chemical composition is displayed in the Science Museum in London. It is interesting to note that at the time of its construction, it was believed that there were two hydrogen bonds between cytosine (C) and guanine (G), rather than three. The DNA double helix model in the Science Museum is therefore inaccurate.

Nucleoside bases have additional roles within the cell

Nucleoside bases can have roles within the cell in addition to their inclusion in DNA and RNA. For instance, adenines make up **ATP**, **adenosine triphosphate**, which stores and carries energy for cells (**Figure 1.7**). ATP stores and releases energy by means of a string of three phosphates attached to the sugar, thus the name triphosphate. Loss of the last of the three phosphates releases energy for the cell from the breaking of the high energy bond between the phosphates. The resulting molecule after the loss of one phosphate is **ADP**, **adenosine diphosphate**, because there are now two phosphates attached to the sugar. Loss of the second phosphate to release more energy produces **AMP**, **adenosine monophosphate**, with just a single phosphate. The energy generated in the cell, from metabolism, is stored in part by adding phosphates back to produce ATP again.

The structures of the purine nucleosides guanine and adenine include a double ring, whereas cytosine, thymine, and uracil have a single ring as pyrimidines. Cell signaling can involve purine-based molecules cGMP, cAMP, and ppGpp. Some nucleosides are also enzyme cofactors, including coenzyme A, FAD, FMN, NAD, and NADP+.

Figure 1.6 The DNA double helix on the £2 coin. In 2003, £2 coins were minted in the United Kingdom to commemorate the description of the structure of the DNA double helix.

Artificial nucleic acids have been created in the laboratory

Artificial nucleic acids are known as **xeno nucleic acids** or **XNA**. In these artificial forms, the sugar is neither ribose nor deoxyribose and is instead an alternative sugar. The XNAs that have been produced include FANA (containing 2′-fluoroarabinose), ANA (containing arabinose), TNA (containing threose), CeNA (containing cyclohexene), GNA (containing glycol), and HNA (containing hexitol). A locked nucleic acid (LNA) has also been generated in the laboratory and shows promise for use in biotechnology applications. In LNA there is a ribose, as in RNA, however, the oxygens in the sugar are connected by an additional covalent bond bridge. This effectively locks the sugar into this form, rather than other forms the ribose can take in nature. PNA, peptide nucleic acids, have been made, which have a pseudo-peptide containing backbone. There are also artificial nucleotides or unnatural base pair (UBP) combinations, including d5SICS and dNaM.

Figure 1.7 The chemical structure of ATP and ADP. Notice that in the name triphosphate is the clue to the structure having three phosphates. The addition of water (H_2O), releases one of these phosphates, generating the structure adenosine diphosphate with two phosphates. This process produces a free phosphate and releases energy.

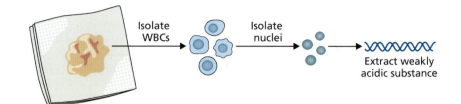

Figure 1.9 Miescher's process that first isolated DNA. Friedrich Miescher, a chemist from the mid-1800s, interested in the components within white blood cells, collected bandages covered in pus. He was able to isolate from the bandages intact white blood cells and from these isolate the nuclei from within the cells. Miescher extracted from the nuclei a weakly acidic substance, which later became known as DNA.

Figure 1.8 The man who discovered DNA, Friedrich Miescher (1844–1895). Working with bandages containing human pus, Friedrich Miescher was the first person to isolate and describe DNA. (From Wellcome Collection. Published under CC BY 4.0.)

DNA was discovered in 1869 and identified as the genetic material 75 years later

Contrary to popular culture, DNA was not discovered by James Watson and Francis Crick in 1953. Swiss chemist Friedrich Miescher (**Figure 1.8**) discovered DNA in 1869, 84 years before Watson and Crick.

Working on the components within white blood cells, leukocytes, Miescher discovered a substance that he called nuclein within the nuclei of these cells. Miescher was attempting to isolate white blood cell proteins. From a local hospital, he obtained bandages containing fresh pus from wounds (**Figure 1.9**). This pus, in the era before antibiotics, would have been an abundant source of fairly pure leukocytes.

Miescher found a weakly acidic substance in the nuclei that had a high phosphorous content and that did not dissolve in any solution that would have dissolved proteins. He therefore knew that this was not a protein and something newly discovered. Nuclein became called nucleic acid and is now called deoxyribonucleic acid (DNA).

The first person to visualize DNA in a living cell was Walther Flemming. In 1879, through microscopy of nuclei, Flemming revealed threadlike structures that are the chromosomes. Flemming investigated how these chromatids, as he called them, separate during cell division. In 1881, Albrechet Kossel showed that Miescher's nuclein was a nucleic acid and named it deoxyribonucleic acid. Kossel isolated and identified the five bases of nucleic acids, A, T, G, C, and U. At the time, isolation of nucleic acids would not have generated pure DNA, therefore it is not unexpected that Kossel found uracil in his samples, as they would have also included RNA. In the early 1900s, Gregor Mendel's experiments in cross breeding plants and observations of the inheritance of traits from the mid-1800s were re-discovered. In the late 1900s, two scientists working independently on different model systems, Walter Sutton and Theodor Boveri suggested that Mendel's units of inheritance were contained within chromosomes.

Phoebus Levene (**Figure 1.10**) was the first to discover some of the key properties of Miescher's nuclein. Publishing in 1919, Levene described the three components of nucleic acids: phosphate; sugar; and base, providing further definition and detail on the three parts that make up DNA, beyond Kossel's identifications of the bases. Levene was also the first to discover the sugar component of both RNA (ribose) and DNA (deoxyribose). By breaking down yeast nucleic acids, Levene discovered that nucleic acids are made of a series of nucleotides made up of phosphates, sugars, and nitrogenous bases. At the time, DNA was not yet recognized as the vehicle for genetics and inheritance, yet Levene studied the molecule in depth, identifying how the three components of nucleic acids form together into nucleotides, which in turn join into chains.

In 1928, Frederick Griffith's investigations revealed important insight into the transfer of genetic traits in bacteria. Using two different morphologies of *Streptococcus pneumoniae*, one that was virulent and killed mice and one that was not, Griffith explored the transfer of genetic traits in bacteria. The virulent version of the bacteria had a smooth morphology due to its bacterial capsule. The non-virulent version of *S. pneumoniae* had a rough appearance when grown on media plates and had no capsule. When the virulent bacteria were heat killed, they did not kill mice. However, when the heat-killed virulent bacteria were injected into mice along with the non-virulent bacteria, the mice died. Bacteria could be isolated from the dead mice and these showed that the isolated

Figure 1.10 Phoebus Levene (1869–1940) uncovered the molecular make-up of nucleic acids. Levene discovered that nucleic acids are made up of phosphates, sugars, and bases and that DNA contains deoxyribose as its sugar and RNA contains ribose as its sugar.

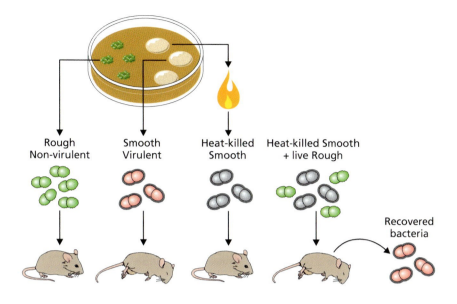

Figure 1.11 The isolation of the transforming principle by Frederick Griffiths. In 1928, Frederick Griffith conducted experiments with *S. pneumoniae*. Injecting rough, non-virulent bacteria into mice did not kill the mice. Injecting smooth, virulent bacteria into mice killed the mice. If, however, the smooth, virulent bacteria were first heat killed and then injected into mice, they lived. This showed that the virulent bacteria could be rendered safe by heat killing them. In the pivotal last part of the experiment, the rough, non-virulent bacteria were injected into mice along with the smooth, virulent bacteria, which had been heat killed. These mice died and bacteria were recovered that were smooth in appearance. The trait of being smooth and virulent was transferred from the dead bacteria to the living rough, non-virulent bacteria.

bacteria were now smooth in appearance, although the only living bacteria injected into the mice in the experiment were the rough non-virulent *S. pneumoniae*. The living rough, non-virulent bacteria had been transformed by something within the heat killed bacteria into smooth, virulent *S. pneumoniae* that killed the mice (**Figure 1.11**). Griffith had identified what he called the transforming principle, a substance capable of transferring the traits of one cell to another. At the time, the prevalent theory was that proteins were the genetic material, because DNA was believed to be too simplistic a molecule to be the blueprint for life. The earliest experiments on the genetic material and the transfer of genetic traits was done in bacteria, paving the way for advances in understanding bacterial genetics and genomics.

Oswald Avery, Colin MacLeod, and Maclyn McCarty, in 1944, showed that the hereditary units, such as those described by Mendel, are DNA, not protein. Using a similar experimental design as had been employed by Griffith, the transfer of traits between *S. pneumoniae* was investigated in an attempt to answer whether DNA or protein or something else within the cell was responsible for the virulence characteristic of the bacteria. In the Avery, MacLeod, McCarty experiments, the smooth, virulent bacteria were heat killed and then subjected to enzymatic digestion. The enzymes used would selectively degrade different components of the heat-killed cells: carbohydrase to degrade carbohydrates; lipase to degrade lipids; protease to degrade proteins; RNase to degrade RNA; and DNase to degrade DNA. The end products of these enzymatic digestions were combined with rough, non-virulent *S. pneumoniae* and the resulting transformants were plated onto agar media. In all cases except one, both rough, non-virulent bacteria and smooth, virulent bacteria were recovered, the latter having been generated through the acquisition of the genetic traits for the capsule, and thus virulence, from the cellular material of the dead, heat-killed, enzymatically digested bacteria. However, when the enzyme involved was DNase, no smooth, virulent bacteria were recovered. This demonstrated that DNA is required for the transfer of genetic traits, therefore DNA is the genetic material (**Figure 1.12**).

The first X-ray images of DNA were taken in 1937 with the structure finally solved in 1953

In 1937, William Astbury, who had trained with the scientists who invented the methodology, was the first to use x-ray crystallography to show that DNA had a regular structure. The attribution of a double helix structure uncovered by x-ray crystallography was later claimed by Rosalind Franklin in 1953, who was able to produce high-resolution images showing the phosphates on the outside. However, the first x-rays of DNA by Astbury were previous to this and already suggested a regular structure to the nucleic acid.

Figure 1.16 Formation of a dinucleotide. The OH of a phosphate group on one nucleotide and the OH group on the 3′ carbon of the other nucleotide are the site of the bond that creates a dinucleotide. The formation of the bond liberates two hydrogens and one oxygen (red), which together form a single water molecule, H_2O (red).

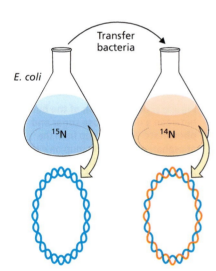

Figure 1.17 The Meselson–Stahl experiment using *E. coli* in media containing heavy nitrogen (^{15}N). *E. coli* were grown in media containing ^{15}N (blue flask). Once the bacteria had fully integrated ^{15}N into its DNA (circular double helix with two blue strands), the bacteria were collected and transferred into fresh media with ^{14}N (orange flask). The results showed that the copied DNA was half from the original DNA, labeled with ^{15}N (blue strand), and half newly synthesized DNA, which used the normal ^{14}N (orange strand). Therefore, we know that when DNA is copied in replication, the two new double helices are made. In each, one strand originates from the starting DNA and one strand is newly synthesized.

one joins to the 3′ carbon of the other in a phosphodiester bond and releases H_2O (**Figure 1.16**). An **oligonucleotide** is a few nucleotides, joined together in a strand. In molecular biology, the term oligonucleotide is used to describe short synthetic pieces of DNA.

The bases across from one another in the DNA double helix are joined by hydrogen bonds. There are two hydrogen bonds between adenine (A) and thymine (T) and three hydrogen bonds between guanine (G) and cytosine (C), making the bond stronger between guanine and cytosine.

When DNA is heated, these hydrogen bonds break, first at the adenine and thymine bonds and then at the guanine and cytosine bonds. Therefore, the sequence influences the melting temperature that separates the two strands. Sequences with more guanines and cytosines have higher melting temperatures than those with many adenines and thymines. We will see later in this book how this has been exploited in molecular biology and how some bacterial species have a higher level of guanines and cytosines or a higher content of adenines and thymines.

Production of nucleotides in the cell is essential so that processes such as replication of DNA and transcription of RNA, described later in this book, can happen. Nucleotides are also required for maintaining and repairing DNA in the cell. But if there is an excess of nucleotides, this wastes energy for the cell. The production of nucleotides is therefore tightly regulated to make the right amount of each that is needed.

DNA is copied semi-conservatively every time a cell divides

The copying of DNA by a cell is called replication. This process is essential. If the DNA within the bacterial cell is not copied, then during cell division, where a single bacterial cell is split into two daughter cells, there will not be enough genetic material for both daughter cells.

The division of individual bacterial cells into daughter cells enables a population of bacteria to expand. The population, if it has grown from a single bacterial cell, is clonal, meaning that all of the cells are copies of the original cell. Any variations in the DNA between cells are mutations, which are likely to have been generated during replication, but can occur by other mechanisms, which will be discussed in later chapters. The concept of DNA replication, the process by which DNA is copied, is simple, however, the various components and mechanisms involved require complex coordination of nucleic acids and proteins.

Matthew Meselson and Franklin Stahl conducted experiments that showed in 1958 that DNA is copied in a semi-conservative manner, meaning that one strand of the double helix comes from the original DNA and one strand is newly synthesized. *Escherichia coli* was grown in media containing ^{15}N, rather than the normally found ^{14}N

(**Figure 1.17**). This allowed the incorporation of this heavy nitrogen into the molecules within the cell, including DNA. Once the ^{15}N-containing DNA had been established in the culture, the bacteria were put into media containing normal ^{14}N. When the *E. coli* DNA was replicated in the ^{14}N-containing media, any new nucleotides created would no longer contain the heavy ^{15}N. In this way, it was shown that replication generates a double helix with one original strand, in this experiment labeled with ^{15}N, and one newly synthesized strand, in this experiment containing ^{14}N.

In the process of DNA replication, the hydrogen bonds in a small section of the DNA are broken and the strands come apart. It should be remembered that the rest of the double helix is still together; the entire double helix of the chromosome does not come apart. Free nucleotides in the cell, as **deoxyribonucleoside triphosphates**, are able to hydrogen bond to the now available nucleotides in the DNA (**Figure 1.18**). The three phosphates on the bases to be used for replication are essential. The 3′ OH on the sugar attacks the alpha phosphate, the one farthest in, and displaces the other two phosphates (beta and gamma). This is called nucleophilic attack and produces a phosphodiester bond.

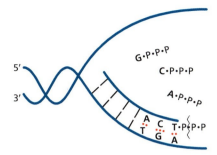

Figure 1.18 Addition of free nucleotides to a DNA strand. Free nucleotides in the cell are available as GTP, CTP, ATP, and TTP, each with three phosphates, deoxyribose, and the base. When added to a DNA strand, two of the phosphates are removed, providing the energy for the formation of the phosphodiester bond between the nucleotide and the DNA backbone.

Replication starts at the origin of replication and requires primers

Replication starts at a place in the bacterial chromosome called the **Origin of Replication**. Helicase recognizes the Origin of Replication and breaks the hydrogen bonds between the DNA bases at this location. The split in the DNA double helix caused by helicase forms what is referred to as the replication fork, a split that progresses along the DNA as it is replicated. To prevent the reformation of the hydrogen bonds, which would bring the DNA strands back together, single-strand binding proteins attach to the strands that were separated by helicase (**Figure 1.19**). Nucleotides are needed for replication to occur, both deoxyribonucleoside triphosphates and also ribonucleoside triphosphates, plus extra ATP to provide energy to fuel the action of the replication proteins.

The major protein involved in DNA replication is an enzyme called **DNA polymerase**. Unfortunately, this enzyme can only extend an existing nucleic acid strand. When the origin separates due to the breaking of hydrogen bonds between bases, there is nothing DNA polymerase can do. DNA polymerase needs there to be some nucleotides on the new strand to get started.

Therefore to start the process of DNA replication, an enzyme called **primase**, which is able to freshly incorporate nucleotides on a new strand, forms a short oligonucleotide of ribonucleic acids. This is an RNA primer, which has been synthesized complementary to the original strand of DNA. With this RNA primer in place, DNA polymerase can add deoxyribonucleotide bases, progressing the synthesis of the new DNA strand (**Figure 1.20**).

Figure 1.19 Helicase opens the DNA double helix to initiate replication at the Origin of Replication. The protein helicase opens the DNA double helix at the Origin of Replication (arrows) by breaking the hydrogen bonds between the nucleotide bases. It progresses toward the replication fork, where the DNA is still together, breaking hydrogen bonds as it goes. In its wake, single-stranded binding proteins coat the single-stranded DNA to prevent the hydrogen bonds reforming.

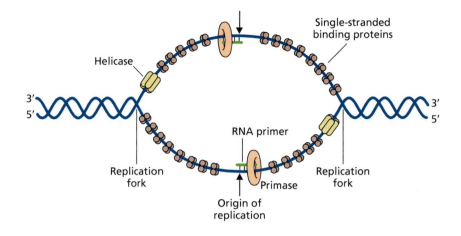

Figure 1.20 Primase incorporates ribonucleic acids to generate a short RNA primer. The enzyme primase initiates the synthesis of the new DNA strands by incorporating ribonucleotides, making a short RNA-based primer. These are required because DNA polymerase III can only add on to an established strand, not create a new one, like primase.

DNA polymerase can only add bases in the 5′ to 3′ direction

The new DNA strand is built 5′ to 3′ through the action of DNA polymerase III, a specialist polymerase for generation of DNA copies. It cannot function in the reverse direction, 3′ to 5′. This process works fine for one strand, but not for the other, which is in the opposite orientation. On one strand, replication can progress smoothly toward the **replication fork**, 5′ to 3′, progressing from the point where the original two DNA strands come apart to be copied to where the original strands are still hydrogen bonded to one another. However, on the opposite DNA strand, replication must progress in short bursts that move away from the replication fork, in the 5′ to 3′ direction (**Figure 1.21**).

On the **leading strand**, going toward the replication fork, primase need only initiate the copying of the DNA strand once through the creation of a single RNA oligonucleotide stretch. On the **lagging strand**, however, RNA primers are made repeatedly, as are the short sections of DNA that connect to them. This enables DNA polymerase to add new nucleotides in a 5′ to 3′ direction, even though the replication fork is moving in the opposite direction. Lagging strand fragments are about 1,000 bases of DNA until it comes to the next primer. These fragments are called **Okazaki fragments**. The RNA primer is removed by a specialized DNA polymerase called DNA polymerase I, replacing the RNA bases with DNA bases. The gap left behind is filled by DNA polymerase I, which now uses the Okazaki fragment as a primer.

The final bond between these extended fragments, now completely made of DNA, is joined by an enzyme called **ligase**. Ligase covalently joins the 3′ OH on a 3′ nucleotide to the 5′ phosphate adjacent. Although DNA polymerase can add new nucleotides to the growing DNA strand, it cannot join two backbones of nucleotides

Figure 1.21 Deoxyribonucleotides are added in replication by DNA polymerase III. The RNA primers are extended by DNA polymerase III, incorporating complementary deoxyribonucleotides in the 5′ to 3′ direction along the DNA. On the leading strand, which heads toward the replication fork, DNA is replicated continuously. On the lagging strand, DNA is replicated in short fragments that head away from the replication fork.

Figure 1.22 The lagging strand in DNA replication is made of Okazaki fragments. While the DNA polymerase progressing toward the replication fork makes a continuous leading strand, the lagging strand is made in short pieces. These Okazaki fragments start as part RNA primer and part DNA. Primase creates new RNA primers as the lagging strand progresses, providing the means for DNA polymerase III to add complementary DNA bases. The RNA primers are removed and replaced with DNA by DNA polymerase I. The gaps in the backbone between the Okazaki fragments are joined by the ligase enzyme.

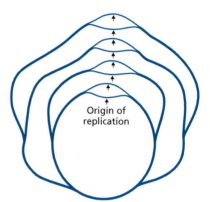

Figure 1.23 Progression of the two replication forks around the bacterial chromosome. At the start of replication of most circular bacterial chromosomes, two replication forks are created. As replication progresses, these move away from each other and around the circle of DNA. Another round of replication can begin once the first has moved away from the Origin of Replication, meaning that there can be several copies of the Origin of Replication and the nearby genes in the cell at any one time.

that have been incorporated at either end of a fragment. This final bond must be made by the separate ligase enzyme.

In bacteria, two replication forks are made at the Origin of Replication, one traveling in each direction around the circular chromosome (**Figure 1.22**). Therefore, one newly formed strand is the leading strand going toward one replication fork and is also simultaneously the lagging strand going toward the other replication fork. As the new strands are synthesized, the replication forks move around the circular chromosome. This process creates stress in the structure of the chromosome. The double helix structure of the DNA must be untwisted to allow the fork to continue to progress. This is accomplished by a topoisomerase II, known as **DNA gyrase**.

Once the replication forks have moved away from the Origin of Replication, another round of replication can begin. This means that it is not necessary to wait until the complete chromosome is replicated before another replication cycle is begun. Therefore, the DNA at the Origin of Replication is often present in more than one, indeed multiple copies, which may influence the expression and function of any associated genes (**Figure 1.23**).

Bacterial DNA can occur in several different forms

In general, bacteria have one circular chromosome that replicates bidirectionally. The **chromosome** therefore contains all of the genetic material for the bacteria to survive and reproduce. During DNA replication it is this chromosome that is copied, so that the genetic traits can be passed to both daughter cells during bacterial cellular division. For the majority of bacteria, the majority of essential genetic features are contained in this one circular chromosome (**Figure 1.24**). There are exceptions to this description, of course, and it should not be assumed that all bacteria are the same. The discovery of bacteria with multiple chromosomes, linear chromosomes, and essential genes outside of what is believed to be the chromosome has caused some review and reassessment of the genetic material within some species of bacteria.

All of the genetic material within the bacterial cell contained on the chromosomes is considered to be the **genome**. Regardless of the number of chromosomes involved, the genetic code contained on them is collectively known as the genome. Any accessory DNA that is extra-chromosomal, not on the chromosome, is considered by many as not part of the genome because extra-chromosomal elements can be easily lost or gained by the bacteria. This genetic material is generally on **plasmids**.

Plasmids are small circular extra-chromosomal DNA molecules found in some bacterial cells. In general, plasmids do not contain essential genes, but may carry genes that could be beneficial to the bacterial cell. These extra-chromosomal pieces of DNA replicate independently and segregate themselves into the daughter cells during division independently of the chromosome. Plasmids can be of varying sizes, but are generally much smaller than the chromosome. Plasmids can contribute genes to the bacterial repertoire that are in addition to those found in the genome.

Figure 1.24 The DNA within a bacterial cell. Most bacterial cells contain a circular chromosome. This is illustrated here as a large circle within the cell (left). Actually, the chromosome is much larger than the cell itself and is coiled up tightly within the cell, so what is shown here is representative of a chromosome and not to scale with the rest of the cell. In addition, the bacterial cell may contain one (center) or more plasmids (right) that are present in single or multiple copies.

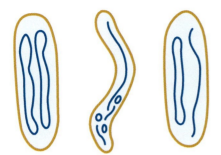

Figure 1.25 Different chromosomal arrangements in bacterial cells. In *Rhodobacter sphaeroides* and other bacteria, two circular chromosomes are present in the cell (left). These are here illustrated as large circles within the bacterial cell, however, the chromosomes are actually much larger than the cell and are tightly wound to fit within the cell. *Borrelia burgdorferi* contains a linear chromosome, as well as several linear and circular plasmids (center). In *Agrobacterium tumefaciens*, both a linear chromosome and a circular chromosome are present (right).

As with all "rules" in biology, there are, of course, exceptions, where it has been found that essential genes are present on plasmids, where bacteria have more than one chromosome, and where bacteria contain linear chromosomes (**Figure 1.25**). Two circular chromosomes have been found in *Rhodobacter sphaeroides*, *Brucella* spp., and *Vibrio* spp. *Paracoccus denitrificans* has three circular chromosomes. *Borrelia burgdorferi* was the first bacteria discovered to contain a linear chromosome, as do *Streptomyces* spp. There is one linear chromosome and one circular chromosome in *Agrobacterium tumefaciens*. Therefore, caution should be taken when assessing the DNA within a bacterial cell and making assumptions about the chromosomes, plasmids, and significance of the genetic material.

Additional genetic material that can be found within bacterial cells can come from bacteriophage. Bacteriophages are small, virus-like particles that infect bacterial cells. These can either kill the bacteria outright by causing lysis in the **lytic phase** or they can incorporate their DNA into the genetic material of the bacteria, entering a **lysogenic phase**. When bacteriophage genomes, or fragments thereof, are integrated into bacterial genomes, the bacteriophage genome is then referred to as a **prophage**.

Prophage genomes can remain dormant in bacterial genomes for many generations. It is possible that these prophage genomes will accumulate mutations that prevent the bacteriophage from activating. Inactive prophage sequences mean that no new bacteriophage particles are made and lytic phase is never achieved. The genetic material from the bacteriophage remains in the bacterial genome as a prophage, where the genes may contribute to the biology of the bacteria, or they may not.

After replication, the two chromosomes produced need to be segregated into what will become the new daughter cells. Each daughter cell must have one complete genome to be viable. It is therefore necessary for the bacterial cell to ensure that the chromosomes are moved into the two halves of the cell before division is complete. To accomplish this, partitioning proteins, which have different names in different bacterial species, bind to the chromosome near the Origin of Replication. These proteins then aid in ensuring that the chromosomes are segregated into the daughter cells. In some species this is accomplished through association with the poles of the dividing cells and in others at locations near the division.

The replication of plasmids is independent of the replication of the chromosome(s)

The processes involved in copying the DNA are similar, however, the replication of plasmids is a separate, independent event. An Origin of Replication, just as is seen in the chromosome, is present in the plasmid and from here replication around the circle is initiated. At any one time, one or more complete copies of a plasmid may be present in a bacterial cell. The copy number of a plasmid is dictated by the plasmid itself. Due to their independent replication outside of the chromosome, their potential to vary in copy number, and their relative ease of isolation and manipulation in the laboratory due to their size, plasmids have been exploited by researchers who use molecular biology techniques to better understand biology and to make it work to achieve desired outcomes.

It should be remembered, when studying the genetics and genomics of bacteria that the genome is often in one chromosome, but not necessarily. Likewise, it should be remembered that the chromosome is often circular, although not always. Therefore, when assessing a DNA code, the circle has been broken and the first base of the code was actually preceded by the last base of the genome sequence.

Key points

- The RNA World hypothesis states that the first nucleic acids were RNA.

- Evolution started with the Initial Darwinian Ancestor, an RNA-based life form capable of replicating itself.

- There was a Last Common Ancestor before the bacteria, eukaryote, and archaea lineages emerged.

- DNA and RNA are nucleic acids, made up of sugars, phosphates, and nitrogenous bases.

- RNA has ribose sugars, phosphates, and adenine, cytosine, guanine, and uracil bases.

- DNA has deoxyribose sugars, phosphates, and adenine, cytosine, guanine, and thymine bases.

- DNA forms a double helix with a backbone joined by phosphodiester bonds and a core joined by hydrogen bonds between bases on opposite strands.

- There are two hydrogen bonds between adenine and thymine and three hydrogen bonds between guanine and cytosine.

- The two strands of DNA run in opposite directions, termed antiparallel.

- Free nucleotides, not part of DNA or RNA, have roles in bacterial cells for energy storage, signaling, and as enzyme cofactors.

- DNA was discovered in 1869 by Friedrich Miescher.

- It was 75 years after Miescher, in 1944, that DNA was shown to be the genetic material by Avery, McCarthy, and McCloud.

- Copying of DNA in replication is semiconservative, meaning one strand is from the previous DNA and one strand is new.

- DNA replication requires the action of the helicase enzyme, single-stranded binding proteins, ATP (for energy), DNA polymerase I, DNA polymerase III, ribonucleotide bases, deoxyribonucleotide bases, ligase, and DNA gyrase (topoisomerase II).

- Generally, bacteria have one circular chromosome, although some may have more, and some may have linear chromosomes.

- The genetic material in the chromosomes is the genome.

- In some bacteria, there is extra-chromosomal DNA, plasmids, which can carry additional genetic information.

Terms, questions, and discussions

Key terms

Define the following terms introduced in this chapter. Check your answers using the definitions in the Glossary. These terms are also available as Flashcards on-line.

Adenine	Hydrogen bond	Origin of replication
Adenosine triphosphate	Lagging strand	Plasmid
Bacteriophage	Leading strand	Primase
Chromosome	Ligase	Prophage
Cytosine	Lysogenic phase	Purine
DNA gyrase	Lytic phase	Pyrimidine
DNA polymerase	Nitrogenous base	Replication
Deoxyribonucleic acid	Nucleic acid	Replication fork
Deoxyribose	Nucleoside base	Ribonucleic acid
Dinucleotide	Nucleotide base	Ribose
Genome	Okazaki fragment	Thymine
Guanine	Oligonucleotide	Uracil

Self-study questions

Answer each question using 50–100 words or a table or labeled diagram. Advice on where to find answers to these questions is available on-line.

1. In the RNA World hypothesis, RNA came first, later followed by DNA. Explain the main structural differences between RNA and DNA.

2. Draw the chemical structure of a segment of RNA, including the linkages of the backbone, the structure of the ribose, and at least one base.

3. What are the names of the four nucleotides found in RNA?

4. What is the name of the force that holds together the two strands of the DNA double helix and how does it vary along the length of the DNA?

5. When and by whom was DNA first discovered?

6. What did Griffith's experiment with *Streptococcus pneumoniae* show?

7. How did Avery, McCarty, and MacLeod build on Griffith's findings?

8. The antiparallel strands of DNA are labeled 5′ or 3′. Explain this designation.

9. When a dinucleotide forms, water is released. Draw the chemical structures involved in this process.

10. Draw the key features involved at the initiation of replication of DNA.

11. Replication requires the introduction of short segments of RNA, so that DNA polymerase can add deoxyribonucleotide bases. What protein is involved, what term is used to describe the RNA, and how frequently is the RNA made in replication?

12. Draw the progression of replication from the Origin of Replication toward the replication forks, showing the leading and lagging strands.

Discussion topics

These topics are presented for discussion in study groups, as part of class discussions, or on your own. These questions go beyond what is directly covered in this part of the book. Use the research literature and other reading to explore these topics in more depth. Tips to help prepare for topic discussions are available on-line.

13. The Last Universal Common Ancestor, known also as LUCA, was a lifeform present before the split in evolutionary lineages into bacteria, eukaryotes, and archaea. What common features still exist in the RNA and DNA processes between these three distinct lineages that support a common ancestry?

14. There is a long period of history between the isolation of DNA from pus on bandages to the demonstration that it is the genetic material in *S. pneumoniae* experiments. During that time it was believed that proteins were the genetic material; DNA was too simplistic. Discuss how overcoming long-standing assumptions such as this have contributed to experimental design and use of positive and negative controls today.

15. When thinking of replication it is often easy to think of it as happening once around a circular chromosome. Discuss the ways in which this is not the way replication occurs in a typical bacterial cell, keeping in mind all of the various kinds of DNA that can be present in a bacterial cell and the variety of DNA in different bacterial species.

Further reading

Life originated from RNA with DNA evolving later

Bowman JC, Hud NV, Williams LD. The ribosome challenge to the RNA world. *J Mol Evol*. 2015; 80: 143–61.

Orgel LE, Crick FHC. Anticipating an RNA world some past speculations on the origin of life: Where are they today? *FASEB J*. 1993; 7(1): 238–9.

Poole AM. Getting from an RNA world to modern cells just got a little easier. *BioEssays*. 2006; 28: 105–8.

Poole AM, Horinouchi N, Catchpole RJ, Si D, Hibi M, Tanaka K, Ogawa J. The case for an early biological origin of DNA. *J Mol Evol*. 2014; 79: 204–12.

Yarus M. Getting past the RNAWorld: The initial Darwinian ancestor. *Cold Spring Harb Perspect Biol*. 2011; 3: a003590.

Nucleic acids are made of nucleoside bases attached to a phosphate sugar backbone

Pinheiro VB, Taylor AI, Cozens C, Abramov M, Renders M, Zhang S *et al*. Synthetic genetic polymers capable of heredity and evolution. *Science*. 2012; 336(6079): 341–4.

Watson JD, Crick FH. Genetical implications of the structure of deoxyribonucleic acid. *Nature*. 1953; 171(4361): 964–7.

DNA was discovered in 1869 and identified as the genetic material 75 years later

Avery OT, MacLeod CM, McCarty M. Studies of the chemical nature of the substance inducing transformation of pneumococcal types. Induction of transformation by a desoxyribonucleic acid fraction isolated from *Pneumococcus* Type III. *J Exp Med*. 1944; 79: 137–58.

Dahm R. Discovering DNA: Friedrich Miescher and the early years of nucleic acid research. *Hum Genet*. 2008; 122: 565–81.

Levene PA. The structure of yeast nucleic acid. IV. Ammonia hydrolysis. *J Biol Chem*. 1919; 40: 415–24.

Miescher F. Ueber die chemische Zusammensetzung der Eiterzellen. *Med.-Chem. Unters*. 1871; 4: 441–60.

Miko I. Gregor Mendel and the principles of inheritance. *Nat Educ*. 2008; 1(1): 134.

O'Connor C. Isolating hereditary material: Frederick Griffith, Oswald Avery, Alfred Hershey, and Martha Chase. *Nat Educ*. 2008; 1(1): 105.

Wolf G. Friedrich Miescher: The man who discovered DNA. *Chem Herit*. 2003; 21: 10–1, 37–41.

The first X-ray images of DNA were taken in 1937 with the structure finally solved in 1953

Chargaff E. Chemical specificity of nucleic acids and mechanism of their enzymatic degradation. *Experientia*. 1950; 6: 201–9.

Hershey AD, Chase M. Independent functions of viral protein and nucleic acid in growth of bacteriophage. *J Gen Physiol*. 1952; 36(1): 39–56.

Pray LA. Discovery of DNA structure and function: Watson and Crick. *Nat Educ*. 2008; 1(1): 10.

Watson JD, Crick FHC. A structure for deoxyribose nucleic acid. *Nature*. 1953; 171: 737–8.

DNA is copied semi-conservatively every time a cell divides

Hajduk IV, Rodrigues CD, Harry EJ. Connecting the dots of the bacterial cell cycle: Coordinating chromosome replication and segregation with cell division. *Semin Cell Dev Biol*. 2016; 53: 2–9.

O'Donnell M, Langston L, Stillman B. Principles and concepts of DNA replication in bacteria, archaea, and eukarya. *Cold Spring Harb Perspect Biol*. 2013; 5(7). pii: a010108. doi: 10.1101/cshperspect.a010108.

Bacterial DNA can occur in several different forms

Baril C, Richaud C, Baranton G, Saint-Girons I. Linear chromosome of *Borrelia burgdorferi*. *Res Microbiol*. 1989; 140: 507–16.

Hinnebusch J, Tilly K. Linear plasmids and chromosomes in bacteria. *Mol Microbiol*. 1993; 10(5): 917–22.

Suwanto A, Kaplan S. Physical and genetic mapping of the *Rhodobacter sphaeroides* 2.4.1 Genome: Presence of two unique circular chromosomes. *J Bacteriol*. 1989; 171(11): 5850–9.

Toro E, Shapiro L. Bacterial chromosome organization and segregation. *Cold Spring Harb Perspect Biol*. 2010; 2(2): a000349.

Genes

Genes are features in the DNA that encode proteins

DNA, as we have learned in Chapter 1, is a string of nucleotides in a double helix structure, where the two strands are complementary to one another. The nucleotides are present in the DNA in a specific order, which is referred to as the **sequence**. The sequence of one strand of the DNA double helix is complementary to the sequence on the opposite strand. These sequences of DNA nucleotides, A's, T's, G's, and C's, are interpreted by the cell for various functions. The most widely studied function of the DNA sequence within living cells is in its role as **genes**.

Genes are the units of heredity. They are what constituted the transforming principle in the early studies by Griffith, Avery, McCarty, and MacLeod. Genes carry the features and traits that we see in humans, plants, and other organisms on the planet. Genes accomplish their function as units of heredity by carrying instructions on how to make **proteins**. Through the specific sequence of nucleotides in the DNA, a gene encodes the information necessary to make one or more proteins. As we will see later, proteins are important for the structure, maintenance, metabolism, and other processes of the bacterial cell.

The word gene comes from the Ancient Greek *genos*, and was first used by Wilhelm Johannsen at the beginning of the twentieth century. The related term genetics was used a few years earlier by William Bateson, taking the term from the ancient Greek *genetikos*. Johannsen also applied specific terms to the study of the outward appearance of the organism and the genetics of the organism. We use terms, **genotype** and **phenotype**, to differentiate these concepts.

An individual, specific DNA sequence is the **genotype**, encoding the potential to express a trait (**Figure 2.1**). The genotype of a bacterial cell is the potential to display various traits encoded in the DNA. The genetic make-up of the bacteria determines the different traits and characteristics that the cell is potentially able to display and express. In general, the genotype largely remains the same, even as the displayed features of the bacteria may change as needed for optimal growth and survival. Changes to the DNA can arise, often through mutation, which can alter the genotype and therefore impact upon the potential features that can be expressed.

The **phenotype** is the expression of the genetic traits in the genotype (see Figure 2.1). It is different from the genotype in that it describes the features that are actually expressed. It is not the case that all genes are being used to make the proteins they encode at all times, therefore phenotype reflects only those parts of the DNA sequence that are displaying the features it encodes. Phenotypes may change in response to the environment, nutrient availability, presence of antimicrobials, and other factors. The phenotype is dictated by the potential of the genetic traits encoded in the genotype, but it does not necessarily reveal the precise nature of the genes responsible for the genotype.

As an example of the difference between genotype and phenotype, consider the expression of bacterial flagella. Some bacterial cells possess one or more appendages that enable them to be motile; these are called flagella. Through the spinning action of the flagella, bacteria are able to move through their environment. To be able to express flagella and use them for motility, the bacteria must have the genes for flagella. There are many genes involved and these encode a variety of proteins that must all be displayed in order for the bacterial cell to successfully express flagella.

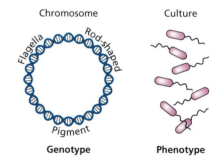

Figure 2.1 The genotype and phenotype of bacteria. The genotype is carried in the bacterial DNA. It is all of the traits that are possible for the organism. For example, the DNA may code for flagella, rod-shaped cells, and production of pigment. The features that are currently displayed by the bacterial cell are the phenotype. For example, these bacteria in culture have flagella, are rod-shaped, and pigmented.

Figure 2.13 The ribosome binds to the mRNA between the start of the transcript and the start of the gene. The ribosome recognizes a sequence between the start of the mRNA transcript and the start of the encoded gene and binds to this region. Specifically, there is a consensus sequence, the Shine–Delgarno sequence that recruits the ribosome to begin translation.

Figure 2.14 The first tRNA recruited during translation is fMet - tRNA$_f^{Met}$. A specific tRNA starts the process of translation. This is fMet-tRNA$_f^{Met}$ and it is recruited straight into the P-site of the ribosome. It contains the anticodon for the initiation codon that is at the start of the gene sequence, which is carried by the mRNA.

Figure 2.12 The three tRNA binding sites of a bacterial ribosome. A cross section shows that there are three locations within the 70S ribosome of bacteria where the tRNAs involved in translation can be found. The first is the A-site, where a tRNA carrying an amino acid enters. This tRNA also carries an anticodon, which matches the mRNA sequence at the A-site. The amino acid on this tRNA will then form a peptide bond to the amino acid(s) attached to the tRNA in the P-site, growing the polypeptide chain by one more amino acid. The tRNA in the E-site has released its amino acid and is ready to exit the process when the ribosome moves along the mRNA to the next codon.

Within the ribosome there are three binding sites for tRNAs: the **Aminoacyl site** (**A-site**); the **Peptidyl site** (**P-site**); and the **Exit site** (**E-site**) (**Figure 2.12**).

Translation initiation

There are four phases of translation: **initiation**; **elongation**; **termination**; and **ribosome recycling**. To initiate translation, the ribosomal subunits come together on the mRNA at a location on the transcript called the **translation initiation region**. This region contains the ribosome binding site, a span of 30 bases that can occur more than once on a bacterial transcript, particularly when that transcript encodes more than one protein (**Figure 2.13**).

Initiation of translation is the most tightly regulated of the phases of translation, limited by the rate of association of the ribosome with the mRNA, a complex that can be prevented through several regulatory mechanisms. It involves the mRNA strand that will be translated, the 50S ribosomal subunit, the 30S ribosomal subunit, three **initiation factor proteins** (IF1, IF2, IF3), and the first tRNA that will be involved in the process. The initiation factor protein IF1 binds to and blocks the A-site of the 30S subunit of the ribosome. Initiation factor protein IF2, the mRNA, and fMet-tRNA$_f^{Met}$ join the 30S subunit. An fMet-tRNA$_f^{Met}$ is a specific type of tRNA that carries a methionine that can be the first amino acid of a peptide chain. A sequence on the mRNA of 5–9 bases, called the **Shine–Dalgarno sequence** or **ribosome binding site**, is 5′ of the initiation codon of the gene, the very first set of three nucleotide bases that signals the start of the polypeptide chain encoded by the gene. In *Escherichia coli* the ribosome binding site has a **consensus sequence** of GGAGG; generally, a sequence similar to this makes up the ribosome binding site. The Shine–Dalgarno sequence on the mRNA hybridizes to the anti-Shine–Dalgarno sequence on the 16S rRNA within the 30S ribosomal subunit. The alignment of the Shine–Dalgarno sequence, aided by the initiation factor protein IF3, puts the initiation codon in the P-site, where the first tRNA, fMet-tRNA$_f^{Met}$, can interact with it (**Figure 2.14**).

The first tRNA used in translation is fMet-tRNA$_f^{Met}$, referred to as the initiator tRNA$_f^{Met}$. It is bound to a special methionine amino acid that has been *N*-formylated. In this way it is different from the tRNAs carrying methionine that may be used later in the process of translation. Elongator tRNA^{Met} also binds to methionine, but is involved in adding this amino acid within the growing peptide chain during the elongation phase of translation, rather than at the start during translation initiation. Each tRNA contains within its RNA sequence a specific three nucleotide sequence that is the reverse complement of codons found in the mRNA. These are **anticodons** and each corresponds to a specific amino acid that is loaded onto the tRNA (see Figure 2.14). The anticodon on the fMet-tRNA$_f^{Met}$ hybridizes to the initiation codon of the mRNA.

Translational elongation

Based on the sequence of the next codon, a tRNA carrying the corresponding anticodon and loaded with the corresponding amino acid enters the A-site to begin the elongation phase of translation. During translation elongation, the mRNA sequence serves as the template for protein production, with the ribosome sliding along the mRNA. With both the P-site and A-site loaded with tRNAs carrying amino acids, the α-amino group of the A-site tRNA attacks the carbonyl group of P-site peptidyl group (**Figure 2.15**). This removes the amino acid from the tRNA in the

Figure 2.15 Peptide bond formation. Peptide bonds are the bonds that join two amino acids together. Peptide bonds form inside the ribosome as a new amino acid is transferred from the tRNA to the growing peptide chain. The peptide bond (red) is formed between the C of the carbonyl group of one amino acid and the N of the α-amino group of the other amino acid. The OH from the carbonyl group of one amino acid and an H from the α-amino group of the other amino acid are released as H_2O during peptide bond formation.

P-site. Depending on how long translation has been happening, this may result in one amino acid bonding to the initial methionine or it may result in one amino acid forming a peptide bond with the growing amino acid chain. The growing peptide chain has now moved to the tRNA in the A-site. The ribosome then slides along the mRNA, resulting in the tRNA from the P-site, with no amino acids attached, moving to the E-site and the tRNA at the A-site, with associated peptide chain, moving to the P-site. This movement of the ribosome along the mRNA results in the A-site becoming free for the next tRNA to enter. Elongation continues at a rate of approximately 20 amino acids per second, with the ribosome sliding along the codons of the mRNA and the tRNAs moving from A-site to P-site to E-site as the peptide chain grows longer. This continues until the ribosome encounters a signal to terminate translation (**Figure 2.16**).

Translational termination and ribosome recycling

The termination phase of translation is brought about when the ribosome encounters a **termination codon**, also called a **stop codon**, which is encoded within the mRNA. Because there is no tRNA with an anticodon that corresponds to the stop codons, instead of a tRNA entering the A-site, a **release factor** binds to the A-site. The presence of the release factor in the A-site triggers the release of the newly synthesized polypeptide chain of amino acids from the ribosome (**Figure 2.17**).

Ribosome recycling is the final phase of translation and is the stage that allows the process to begin again at the next ribosome binding site. Once the peptide chain is released, the ribosomal subunits dissociate and release the mRNA strand. IF3 binds to the 30S subunit, promoting the separation of the subunits and preparing them for the new round of initiation.

Once a ribosome has moved through the initiation phase and the ribosome has progressed sufficiently along the mRNA to make the ribosome binding site available, a second ribosome can join the process of translation from the same mRNA strand. In this way, several peptide chains can be made from one mRNA at the same time. Recall that several mRNAs can be made from the same gene at the same time. In this way, multiple copies of a protein can be quickly and efficiently generated through coupled transcription–translation.

Coupled transcription–translation in bacteria has mRNA being made and used to produce proteins in tandem

Translation takes place in concert with transcription in bacteria. This is referred to as **coupled transcription–translation** and is a feature of bacterial cells. Before the mRNA is finished being made by transcription, it is being interpreted by translation into a protein (**Figure 2.18**). Indeed, more than one protein chain can be made from each mRNA strand. It has been observed that several proteins can be simultaneously generated by translation in the wake of RNA polymerase's action

Figure 2.16 The amino acid chain grows during elongation. As the ribosome progresses during translational elongation, new codons are accessible at the A-site. A tRNA with the corresponding anticodon will enter the A-site, bringing with it the associated amino acid. Present in the P-site is the previously used tRNA, now carrying the whole of the amino acid chain that has been formed. At the E-site is the tRNA previous to the one in the A-site, which has released the amino acids to the tRNA at the P-site and is ready to exit the ribosome.

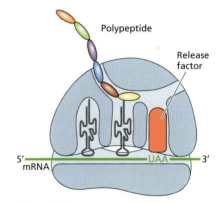

Figure 2.17 Translation ends when a termination codon is reached by the ribosome. Rather than a tRNA bringing another amino acid to add, the presence of a termination codon at the A-site signals the ribosome to stop making the polypeptide. Release factor enters the A-site and triggers the end of translation, where the polypeptide is released from the tRNA in the P-site and both the tRNAs in the P-site and E-site are released. In the final phase, the ribosome dissociates and prepares to start a new round of translation.

Figure 2.18 In bacteria, the processes of transcription and translation are coupled. As the cell is transcribing a gene, the mRNA is being produced. Once enough of the mRNA is made for the ribosome to recognize its binding site, the mRNA can be used for translation. Therefore, proteins are produced from genes as the mRNA is being generated from them.

Figure 2.19 Electron micrograph of coupled transcription–translation. Using high-resolution electron microscopes, it is possible to visualize the process of transcription and translation, demonstrating that in bacteria these processes happen together. Here the DNA strand can be seen, as well as the mRNA that is being transcribed from it. Note that several mRNAs are being made simultaneously. The mRNA is then coated with ribosomes, which translate the mRNA into protein. Note how many ribosomes are attached to the mRNA, simultaneously making several peptide chains from a single mRNA.

in transcription (**Figure 2.19**). Although coupled transcription–translation is often cited as a classic differentiator of bacterial cells and eukaryotic cells, it has been shown to also occur within the nuclei of mammalian cells at sites of transcription "factories," where polypeptides are also made. Coupled transcription–translation in bacteria was clearly demonstrated in the 1970s from electron microscopy images of *E. coli*, showing transcription and translation, including several events of translation occurring off several transcripts originating from the DNA (see Figure 2.19).

Bacteria can have more than one gene on an mRNA strand and form operons

Thus far, we have considered that a transcript encodes a single gene. This is not always the case. In bacteria, more than one gene can be carried on a single mRNA transcript. Therefore, two or more different proteins can be translated from the same transcript. Each protein encoding portion of the mRNA will have its own ribosome binding site, with initiation, elongation, termination, and ribosome recycling occurring separately for each of the protein encoding genes carried by the message. These are referred to as **polycistronic units**, sets of genes that are transcribed together (**Figure 2.20**).

In some cases, differences in the promoters used or other signals in the sequence may mean that at times a gene is part of a polycistronic unit and at other times it is transcribed into mRNA on its own. Bacterial mRNA can be polycistronic, but does not have to be. Genes that are together in a piece of polycistronic mRNA may be an **operon**.

An operon is a region of DNA where a set of genes with related functions are controlled by a common regulatory element and as such generate an mRNA transcript containing these genes. Operons have distinct elements that control the transcription of the genes. This can ensure that a set of genes of related function are expressed in equivalent numbers and/or at the same time, being derived from the same mRNA strand. In cases where a gene can be part of a polycistronic unit or can be transcribed separately from a second promoter, this gene is at times part of the controlled operon and at other times not.

Figure 2.20 Genes as polycistronic units. Groups of genes that are adjacent to one another can be transcribed together when they are part of a polycistronic unit. These are two or more genes that are transcribed into mRNA from the same promoter, generating one mRNA that encodes more than one gene.

Open reading frames are regions of the DNA between termination codons

DNA contains nucleotide bases arranged to encode proteins, RNAs, and to provide genetic signals to the cell. The most common function of the DNA in bacteria is to encode proteins. To do so, a transcribed strand of mRNA must be translated into a polypeptide chain of amino acids that make a protein. The three nucleotide codes within the mRNA corresponds to the amino acids of the protein it encodes, making the codon.

There are 61 tRNAs with different anticodons, sequences within the tRNA that will hybridize to the codon on the mRNA. Three codons have no tRNA with complementary anticodons. In translation, the ribosome will stop adding amino acids to the growing peptide chain when it reaches a codon for which there is no corresponding amino acid. These triplet codes of nucleotides are called termination codons. Each termination codon has a nucleotide sequence in the DNA, which is transcribed into the mRNA, and is known by a semi-precious gem stone name: TAG stop codon in the DNA, UAG in the mRNA, amber; TGA stop codon in the DNA, UGA in the mRNA, opal; and TAA stop codon in the DNA, UAA in the mRNA, ocher. These termination codons signal the release factor to enter the ribosome at the A-site, terminating translation.

There is no tRNA with an anticodon for UAG, UGA, or UAA, however most rules have exceptions, including this one. The exceptions will be explored in Chapter 7. Except in rare circumstances, the presence of a TAG, TGA, or TAA in the DNA will signal the termination of translation. Therefore, it is possible to identify regions containing genes, which will be translated into long polypeptide chains by finding areas of the DNA sequence without stop codons. DNA is double stranded, with a **positive strand** running in one direction and the **negative strand** in the opposite direction along the double helix (**Figure 2.21**). The designation positive strand refers to the strand of DNA encoding a gene, also known as the **sense strand**, while the negative strand is the **antisense strand** containing the complementary bases. These designations are appropriate when the gene is known, but it may be that when trying to identify regions with potential genes, the genes are on either strand. At the genomic level (Chapter 3) these designations do not apply, because genes are found on both strands.

Each codon is made of three bases. To read a codon code, three bases are read together, then the next three, then the next. A different codon code can be read by starting one base farther along and a third different codon code using the base after that. Therefore, there are three different codon codes that can be read on each of the two strands of DNA. These are referred to as **reading frames**. In order for a protein to be made, the reading frame must not contain any stop codons, otherwise these would prematurely signal the end of translation of the protein. The area between two stop codons in the same frame is called the **open reading frame** or **ORF**. Most ORFs are very short and do not encode proteins. Long ORFs, stretches of sequence where there are many amino acid codons but no termination codons, are likely locations for genes (see Figure 2.19).

Figure 2.21   DNA has six reading frames. There are six reading frames in DNA, with three on the forward strand and three on the reverse. Reading frames are the sets of three nucleotides that make up a codon. There are therefore three different combinations of bases on the forward strand and three on the reverse strand.

Figure 2.22 An ORF that is also a CDS.
An open reading frame (ORF) is a reading frame in a section of DNA where there are no stop codons. When there is a start codon early in the ORF, then this feature is a coding sequence (CDS). Note the ATG start codon present immediately after the TGA stop codon at the start of the sequence.

To be a gene, however, an open reading frame must have a start codon as well as having a stretch of amino acids before reaching a stop codon. An ORF that has a start codon at the beginning of the frame is called a **coding sequence** (**CDS**) (**Figure 2.22**). There are three possible initiation codons, however they are not all used in equal proportions and not all codons that carry the same sequence can be taken as the start of translation. The most abundant initiation codon is the ATG start codon, which is decoded to *N*-formylated methionine at the start of the amino acid string. This is present in the mRNA transcript as AUG and is seen in 90% of the coding regions in *E. coli*. The second most abundant initiation codon is the GTG start codon. This is present in the mRNA transcript GUG and decodes to a *N*-formylated methionine. It is the start codon for 8% of the coding regions in *E. coli*. The most rare start codon is the TTG initiation codon, present in the mRNA transcript as the sequence UUG. Like GTG, this also decodes to a *N*-formylated methionine, therefore all polypeptide chains start with *N*-formylated methionine as they are generated on the ribosome.

There are genes that contain what could be interpreted as more than one initiation codon. They may have a TTG codon sequence that is then followed later in the open reading frame by an ATG codon. Which codon is used as the initiation codon by the translation machinery is determined by the location of the ribosome binding site, responsible for initiation of translation. It is therefore important to remember the relative abundance of each initiation codon when determining the location of a CDS. If there is a GTG or TTG present, this may be the initiation codon, but this choice over an ATG should be supported by the location of a ribosome binding site or, ideally, experimental evidence.

When a CDS is a gene

When identified in the DNA sequence, or in the mRNA sequence, a region with both an initiation and termination codon is a coding sequence (CDS). Only experimental evidence showing the expression of the CDS's encoded protein can classify it as a gene. Quite often, evidence of expression of the protein encoded by the CDS is obtained in one strain of one bacterial species and then this information is extrapolated to other strains and even other species of bacteria, provided there is sufficient similarity between the sequences. This similarity may be at the DNA level, where two CDSs share the same nucleotides in many of the same locations. This is **DNA identity** and is a measure of the **homology** of the sequences, or the degree to which they are the same. In some cases, the similarity between two features is most apparent at the protein level of the sequence. In this case, we are comparing the amino acid similarity. There are two measures of amino acid homology, the identity and the similarity. **Amino acid identity**, similarly to DNA identity, is a comparison of the exact sequence. The **amino acid similarity** takes into account the similar nature of some amino acids, due to their charge or interaction with other molecules.

To determine the amino acid identity and similarity from the information in a DNA sequence, the nucleotide sequence must be translated into an amino acid sequence. In this case, it is not bacterial translation, using ribosomes and tRNAs that translate the DNA, via mRNA, into the amino acid strand. Instead, a **codon table** can be used to determine what would happen in bacterial translation, and therefore artificially translate the nucleotide sequence into an amino acid sequence. There are six reading frames in a strand of DNA that can be translated, three on the positive strand and three on the negative strand (**Figure 2.23**). The amino acid sequence amongst these will be the one that is framed by the start and stop codons, the initiation and termination codons.

The codons between the start and stop codons within the CDS, the three-letter codes, one following another, encode the amino acids of the corresponding protein. To determine the amino acid sequence, a codon table is used. This is a table

Figure 2.23 The area between two stop codons in a reading frame is an open reading frame (ORF). The six reading frames of DNA are composed of codons. Some of these codons encode amino acids and some signal the termination of translation, a stop codon. The region between two stop codons in a reading frame is referred to as an open reading frame. A short ORF is shown here in reading frame 1, where a short segment of DNA has no stop codons in this frame.

containing all of the possible combinations of three nucleotide bases and indicates the amino acid to which they correspond or highlights that the code is a termination codon. Often, there is more than one codon that encodes a particular amino acid. This is shown in the codon table, where several codons correspond to the same amino acid. For some of these, there is what is referred to as a **third base wobble**. This means that the first two bases are the key to deciphering the amino acid and the third base in a codon can be different bases, yet still encoding the same amino acid (**Table 2.1**).

TABLE 2.1 Codon Table

First base	Codon	Amino acid	Codon	Amino acid	Codon	Amino acid	Codon	Amino acid	Third base
T	TTT	Phe (F)	TCT	Ser (S)	TAT	Tyr (Y)	TGT	Cys (C)	T
	TTC	Phe (F)	TCC	Ser (S)	TAC	Tyr (Y)	TGC	Cys (C)	C
	TTA	Leu (L)	TCA	Ser (S)	TAA	Stop	TGA	Stop	A
	TTG	Leu (L)	TCG	Ser (S)	TAG	Stop	TGG	Trp (W)	G
C	CTT	Leu (L)	CCT	Pro (P)	CAT	His (H)	CGT	Arg (R)	T
	CTC	Leu (L)	CCC	Pro (P)	CAC	His (H)	CGC	Arg (R)	C
	CTA	Leu (L)	CCA	Pro (P)	CAA	Gln (Q)	CGA	Arg (R)	A
	CTG	Leu (L)	CCG	Pro (P)	CAG	Gln (Q)	CGG	Arg (R)	G
A	ATT	Ile (I)	ACT	Thr (T)	AAT	Asn (N)	AGT	Ser (S)	T
	ATC	Ile (I)	ACC	Thr (T)	AAC	Asn (N)	AGC	Ser (S)	C
	ATA	Ile (I)	ACA	Thr (T)	AAA	Lys (K)	AGA	Arg (R)	A
	ATG	Met (M)	ACG	Thr (T)	AAG	Lys (K)	AGG	Arg (R)	G
G	GTT	Val (V)	GCT	Ala (A)	GAT	Asp (D)	GGT	Gly (G)	T
	GTC	Val (V)	GCC	Ala (A)	GAC	Asp (D)	GGC	Gly (G)	C
	GTA	Val (V)	GCA	Ala (A)	GAA	Glu (E)	GGA	Gly (G)	A
	GTG	Val (V)	GCG	Ala (A)	GAG	Glu (E)	GGG	Gly (G)	G
Second base →		T		C		A		G	

Note: This table shows the amino acids that are encoded by the various three base codon combinations, as well as those for translational termination. The DNA version of the code is shown there, including T's. During translation mRNA is used, therefore the T's would have been transcribed into U's. Both the three-letter amino acid code and the single-letter amino acid code are shown.

Figure 2.24 Single base changes may change the amino acid sequence. Single-nucleotide polymorphisms (SNPs) are changes in a single base of DNA. Depending on the position of the change in the codon, this may change the encoded amino acid, or it may not. As in the codon table (Table 2.1), the same amino acid can be encoded by more than one codon. These differ by one or two bases. As shown, when the bases change, shown in orange on the bottom DNA sequence, the amino acids stay the same in some cases and change in others. SNPs can also affect stop codons.

Mutations that change the DNA sequence may not change the encoded amino acid. As can be seen in the codon table, there is scope for bases to change and to still encode the same amino acid. When describing mutations, such as **single-nucleotide polymorphisms** (**SNPs**), the terms **synonymous** and **non-synonymous** are often used. The SNP has changed just one base in a region of DNA. If that one base change is positioned within a codon such that it still encodes the same amino acid, then this is a synonymous change. Although the DNA code has altered, the codon code has not and the resulting protein will carry the same amino acid sequence. If, however, the base that has been changed does influence the amino acid sequence, causing a change in the codon that means a different amino acid will be placed in the protein during translation, then this is called non-synonymous. The SNP change will have altered the DNA sequence in such a way that the amino acid sequence is also altered (**Figure 2.24**). This may have consequences for the function of the protein. Later in this book events that give rise to SNPs are described, as well as methods that can be used to introduce them.

Although a synonymous SNP will not alter the sequence of the encoded protein, it may alter the rate at which the protein is made. Not all tRNAs are generated within the cell at the same level. Some tRNAs are very rare within the bacterial cell. The use of these codons is selected against evolutionarily; generally, the bacterial genome sequence will favor some codons over others based on the availability of the corresponding tRNAs. If a nonstandard codon that is not commonly found in the species is present within a gene, it may cause a pause in the translation process, as the ribosome awaits the arrival of the correct, rare tRNA. This can result in fewer proteins being made than could be made with a standard codon in this place.

Expression of genes is controlled – not all genes are on all of the time

Promoters are regions of the DNA sequence where transcription is initiated to generate mRNA. These regions are near the start point of transcription in bacteria, generally located just 5′ of the transcriptional start site. Promoters are also on the same strand of the DNA double helix as the start point of transcription. The directionality of the −10 and −35 sequences in the promoter dictate the direction of RNA polymerase transcription. The strand containing the promoter and the start point of transcription is referred to as the sense strand. The promoter is where RNA polymerase begins its work in transcription.

The promoter region is where all transcription initiation factors and RNA polymerase bind to specific DNA sequences to initiate transcription. In addition, other transcription factors may bind to the promoter region and modify transcript generation.

All bacterial promoters must have an RNA polymerase binding site, which is generally the −10 and −35 sequences that are present 5′ of the gene's transcriptional start site. The **−10 promoter sequence** is also called the **Pribnow box**. The sequence of the −10 is usually similar to the consensus sequence TATAAT. The **−35 promoter**

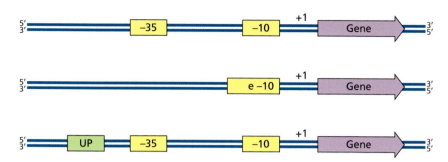

Figure 2.25 The features of a promoter region. Promoter regions frequently include the −10 and −35 sequences that are recognized by the σ^{70} subunit of RNA polymerase. In some instances, an extended −10 region (e − 10) may be present, which compensates for the absence of a −35. An UP element can also be present upstream of the −35, which enhances RNA polymerase binding.

sequence is usually similar to the consensus sequence TTGACA, although this sequence is less conserved than that of the −10 sequence.

There are, however, some promoters that completely lack the −35 portion of the promoter region and instead have an **extended −10 region**. This extension to the −10 sequence compensates for the absence of the −35 region and allows RNA polymerase to bind and initiate transcription.

Other promoters may contain what is referred to as an **UP element**. When it is present, the UP element is 5′ of −35, therefore farther away from the gene than the rest of the promoter (**Figure 2.25**). This additional sequence enhances transcription of the associated gene(s) due to it imparting better binding of the alpha subunit of RNA polymerase. This enhancement efficiently recruits RNA polymerase to the UP element containing promoter and increases the level of transcription.

Sigma factors are responsible for RNA polymerase promoter recognition

The specificity of RNA polymerase for the promoter region sequence is determined by the enzyme's **sigma subunit** (σ). This is one part of RNA polymerase; the sigma factor and the other subunit components of RNA polymerase make up the **holoenzyme** (**Figure 2.26**). Not all RNA polymerase within the bacterial cell will contain identical σ subunits. By changing the σ subunit, the specificity of RNA polymerase for the promoter region sequences is changed.

The σ^{70} (sigma 70) subunit is the primary sigma factor in bacterial cells, responsible for the majority of the genes and those housekeeping genes essential for life. Other sigma factors are responsible for other functions in the cell. Unlike the primary sigma factor, which is in abundance in the cell, the production of these alternative sigma factors is tightly regulated, thus ensuring controlled regulation of the genes associated with the promoters they recognize. Sigma factors enable the simultaneous expression and regulation of many genes across the bacterial genome.

The σ^{54} subunit, for example, is involved in the nitrogen response and therefore controls the relevant genes that possess a different promoter sequence, recognized by σ^{54}. Alternative sigma factors are responsible for several important processes in the bacterial cell and can include response to the environment, response to and survival during conditions of stress, or the development of the cell. Bacterial cells are not static. They must change in accordance with their environment and the situations they encounter. These changes can result in the differentiation of the cell. The process of sporulation, whereby some bacteria make spores containing their genetic material that are resistant to stress, includes control of the transcription of sporulation-specific genes by alternative sigma factors.

Different species have different numbers of sigma factors. Some species have tens of different sigma factors, all recognizing a different promoter region sequence. Other species have only a few sigma factors, relying for the most part on σ^{70} and regulating gene expression through other means. It therefore cannot be assumed that a particular bacterial species possesses a particular sigma factor, even when that

RNA polymerase

Figure 2.26 The subunits of RNA polymerase. RNA polymerase is composed of six subunits as the RNA polymerase holoenzyme, with two α subunits, a β subunit, a β' subunit, a ω subunit, and the σ subunit. The presence of the σ subunit completes the holoenzyme. Without the σ subunit, the RNA polymerase core enzyme remains.

sigma factor is present in other species and has been demonstrated to serve a vital role there.

The alternative sigma factors have different binding sites from the consensus -10 and -35 used by σ^{70}. Therefore, when analyzing DNA sequence data, if no -10 and -35 sequences are found 5′ of a gene, it could be that an alternative sigma factor is responsible for initiation of transcription for this gene. Where the sequences of the alternative sigma factor binding sites are known, it can be beneficial to investigate their presence as well. In some cases, the same gene can be transcribed by RNA polymerase holoenzyme carrying a σ^{70} or by one with an alternative sigma factor.

The sigma factor component of the RNA polymerase holoenzyme is only required for the initiation of transcription. Once the transcription process enters elongation, the sigma subunit is shed and is able to initiate transcription again at the same site or at another promoter through association with another RNA polymerase.

Regulatory proteins change the level of transcription

In addition to RNA polymerase, there are proteins that can bind to the promoter regions in bacteria. These DNA-binding proteins are called **regulators**. These proteins can either bind in association with RNA polymerase and enhance its ability to initiate transcription, as activators, or bind to a region of the promoter that excludes the binding of RNA polymerase, repressing the initiation of transcription.

Repressor proteins prevent or reduce the transcription of genes

Repressors are regulatory proteins that bind to specific DNA binding sites in promoter regions and prevent transcription by RNA polymerase, or in some cases significantly reduce transcription of the 3′ gene(s). Some repressors bind to a region that overlaps the -10 and/or -35 regions of the promoter. This prevents RNA polymerase from binding to the promoter (**Figure 2.27**).

Other repressors bind to sites on the DNA that overlap an activator protein binding site. In these situations both proteins, repressor and activator, are trying to bind to the same location on DNA and only one can do so. When it is the repressor that binds, this prevents RNA polymerase binding to the promoter region or in some cases prevents open complex formation necessary for initiation of transcription.

A repressor may also bind to more than one site in association with the gene for which it represses transcription. Once bound, the multiple copies of the repressor protein interact. This can cause bends, loops, or hairpins in the DNA. These topology changes stop RNA polymerase binding, open complex formation, activator binding, or elongation.

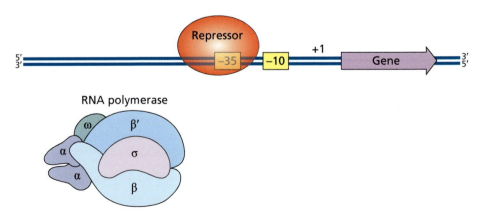

Figure 2.27 The binding of a repressor protein to a promoter region. Repressor proteins are DNA binding proteins that are capable of binding to a promoter region and preventing the binding of RNA polymerase, thereby blocking the initiation of transcription.

Figure 2.28 The binding of an activator can compete with the binding of a repressor.
Activator proteins are DNA binding proteins that assist in the binding of RNA polymerase to a promoter.
When the binding sites for activators and repressors are close to one another or overlapping, only one
of the proteins can bind. When the repressor binds, it prevents RNA polymerase from binding. When the
activator binds, it enhances the binding of RNA polymerase and the initiation of transcription.

Activator proteins contribute to the expression or increased expression of genes

Regulatory proteins that are **activators** are capable of inducing transcription of genes.
Activators, as DNA binding proteins, can bind near or at promoters and thus assist
in the binding of RNA polymerase to the promoter for the initiation of transcription.
In some cases, repressors also bind to the promoter and the two opposing proteins
compete. If the activator binds, the gene(s) are transcribed (**Figure 2.28**).

Activators can also bind to **enhancers**, which are sequence regions that enhance
transcription. Enhancers are regions of DNA that can be 5′ of a gene, within a gene,
3′ of a gene, or distantly. These regions in the DNA need not be in the promoter. The
binding of an activator to an enhancer often results in the DNA bending, enabling
the activator protein to interact with the transcriptional machinery at the promoter
(**Figure 2.29**).

Regulatory RNAs may have an impact upon transcription

There are **small RNAs** made within bacterial cells that are not mRNAs, tRNAs, or
rRNAs. These other RNAs are regulatory RNAs and although many act on the level
of translation there are some, however, that regulate transcription. Most of these act
in **trans**, which means that they are acting on elements that are remote. In some
examples, an antisense regulatory RNA, which has the complementary sequence to
part of the gene, binds to a complementary region on an mRNA leading to premature
transcriptional termination, often between two genes in an operon. Thus, the more
3′ encoded genes of the operon have a lower level of transcription than do those
genes at the 5′ start of the operon.

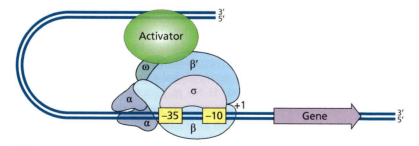

Figure 2.29 Some activators bind distantly from the promoter. In some cases, the activator
protein binds to a region that is upstream of the −35 region by some distance. It contributes to RNA
polymerase binding, enhancing transcription, through the bending of DNA, which brings the activator
together with the RNA polymerase.

**Figure 2.31 Deletion of genes,
duplication of genes, and addition
of horizontally transferred genes.** (a)
During evolution, regions of DNA can be
deleted from the genome, resulting in the
loss of genetic material, potentially whole
genes. (b) During evolution, regions of DNA
can be duplicated, making a second copy
of that segment of DNA. This can result in
the gain of a second copy of some genes.
(c) New genetic material can be added into
the genome when DNA enters the bacteria
from another cell. Once incorporated,
this horizontally transferred material
may result in a gain in function(s) for the
bacteria through acquisition of horizontally
transferred genes.

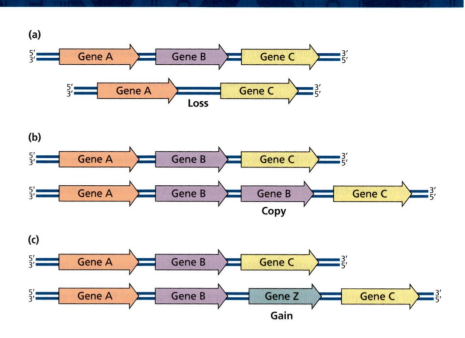

Of course, the gene repertoire within a bacterial cell would be influenced by the environment in which the bacteria grow, availability of nutrients and other resources, and other factors. These conditions can be artificially controlled in the laboratory. It has therefore been hypothesized that there is a minimal bacterial genome, containing only those genes that are absolutely required. Researchers in synthetic biology are working to understand the minimal bacterial genomes through the creation of bacterial chromosomes engineered to contain only specific genes.

Key points

- Genes are those specific regions of the DNA that code for proteins.

- Bacteria can contain a wide variety of genetic traits that make up the organism's genotype.

- The portion of the bacterial cell's genotype, its genetic information, being expressed by the organism is the phenotype.

- The cellular process of bacterial transcription takes the genetic information from DNA and generates messenger RNA based in this sequence.

- There are three phases to bacterial transcription, which are termed initiation, elongation, and termination.

- Transcription is initiated when RNA polymerase holoenzyme binds to the promoter region and begins making mRNA at the start point of transcription.

- The progression of transcription is from the 5′ to 3′ of the DNA template.

- Once transcription enters the elongation phase, the sigma subunit of RNA polymerase is released, then being able to assist in the initiation of transcription with a new holoenzyme.

- Transcription ends when either an intrinsic transcriptional terminator or a Rho-dependent transcriptional terminator is reached.

- Multiple rounds of transcription can occur simultaneously for the same gene.

- Bacterial translation uses the genetic code carried by mRNA to generate proteins with the assistance of tRNAs carrying the amino acids that will make the proteins and rRNA contained within the ribosomes that orchestrate protein production.

- Whilst DNA and RNA are made up of nucleic acids, proteins are made up of amino acids, which come together in polypeptide chains based on the gene sequence.

- During translation, peptide bonds are formed between adjacent amino acids as the peptide chain is generated.

- The gene sequence of nucleotide bases is interpreted in sets of three bases, called codons, for the production of proteins.

- There are four phases of translation that share similar names to the phases of transcription, being initiation, elongation, termination, and ribosome recycling, therefore it is important when discussing initiation to know which process is being described.

- The tRNA molecules contain three base codes called anticodons in their structure, which complement the codons on mRNA.

- The ribosomes in bacterial cells are 70S ribosomes, containing two subunits, the 50S subunit with proteins, 5S rRNA, and 23S rRNA, and the 30S subunit with proteins and 16S rRNA.

- There are three important sites within a ribosome that are essential for the orchestration of protein production, the A-site where tRNAs first enter, the P-site where peptide bonds to the growing peptide chain are made, and the E-site where the uncharged tRNA will exit the ribosome.

- A ribosome, guided by the 16S rRNA within it, will bind to the Shine—Dalgarno sequence that precedes the initiation codon to initiate translation.

- The ribosome slides along the mRNA during translation elongation, providing three base codons for the corresponding tRNA and recruiting amino acids to the growing polypeptide chain.

- Translation stops when the ribosome's A-site comes to a termination codon, signaling the end of translation.

- The final phase of translation releases the polypeptide chain of amino acids that has been created and releases the ribosome so that it can go on to initiate translation elsewhere.

- In bacterial cells, transcription and translation happen together, with the mRNA being made and bound by ribosomes to make proteins before the process of transcription is complete.

- Multiple ribosomes can bind to the same mRNA strand to generate proteins based on the gene simultaneously.

- Genes that are adjacent to one another may be transcribed into the same mRNA, forming a polycistronic unit.

- A set of genes in a polycistronic unit that share the same regulatory controls can be considered to be an operon, where the products of the genes are expressed together in an organized manner.

- There are six potential reading frames of three base codons encoded in the DNA, three on the positive strand and three on the negative strand.

- There are three termination codons, TAG, TGA, and TAA, and the regions within the reading frame that lack termination codons are the open reading frames (ORFs).

- When there is an appropriately positioned initiation codon (ATG, GTG, or TTG) within an open reading frame, this is a coding sequence (CDS) and may encode a protein.

- Not all initiation codons are equally represented in the genetic code of bacteria; GTG is far more rare than ATG and TTG is very uncommon.

- Experimental evidence such as detection of the corresponding mRNA or the production of the encoded protein elevates a CDS from a prediction to a gene.

- Homology is a measure of how much two sequences are the same as each other.

- DNA sequence homology is expressed in identity, whereas protein sequence homology can be expressed as identity, for carrying the same amino acids, or similarity, for carrying amino acids with similar chemical properties.

- The protein sequence can be deduced from the DNA sequence information using a codon table, which identifies the amino acid that corresponds to each codon.

- To initiate transcription the RNA polymerase must bind to the promoter region that sits before the gene, containing specific binding site DNA sequences that are recognized by the subunits of RNA polymerase.

- The sigma subunit of the RNA polymerase holoenzyme is responsible for promoter sequence specificity and therefore governs which genes will be transcribed.

- Bacteria may have many different sigma factors that are used to coordinate expression of distant genes across the DNA.

- Repressor regulatory proteins can prevent or reduce the transcription from a promoter, generally through binding to the promoter region and interfering with the initiation of transcription.

- Activator regulatory proteins can cause or enhance the transcription from a particular promoter through assisting in the binding of RNA polymerase to initiate transcription.

- Although regulatory small RNAs generally exert control on protein expression through influencing translation, there are some that impact upon transcription.

- At some times the sequence within the mRNA itself can impact upon the success of transcription of a gene as a riboswitch.

- Some regulators of transcription act upon more than one promoter region; when these are widespread, they may be referred to as global transcriptional regulators.

- Bacterial species vary in their possession of genes; some are essential and others are beneficial for survival, as accessory genes.

Terms, questions, and discussions

Key terms

Define the following terms introduced in this chapter. Check your answers using the definitions in the Glossary. These terms are also available as Flashcards on-line.

Accessory gene	Peptide bond	Sigma subunit
Activator	Phenotype	Stop codon
Amino acid	Polycistronic units	Termination codon
Anticodon	Promoter region	Transcription
Coding sequence	Protein	Transcription elongation
Codon	RNA polymerase	Transcription initiation
Coupled transcription–translation	Regulator	Transcription termination
Essential gene	Repressor	Transfer RNA
Gene	Ribosomal RNA	Translation
Genotype	Ribosome	Translation elongation
Homology	Ribosome binding site	Translation initiation
Messenger RNA	Sequence	Translation termination
Open reading frame	Shine–Dalgarno sequence	

Self-study questions

Answer each question using 50–100 words or a table or labeled diagram. Advice on where to find answers to these questions is available on-line.

1. What is the specific term for a sequence of DNA that encodes a protein and what features distinguish protein-encoding sequences of DNA from other sequences?

2. For proteins to be made from DNA, first an RNA template has to be made. What kind of RNA is this, what is the name of the process that generates it, and what are the three stages in the process of it being made?

3. Ribosomes are made up of protein and RNA. What kind of RNA is in a ribosome and what are the designations of the different RNAs in a bacterial ribosome?

4. Draw the ribosome in cross-section, showing the binding sites, RNA channel, and channel for the growing peptide chain.

5. What is the name of the process where a sequence of RNA becomes a protein and what are the three stages in the process of the protein being made?

6. What are the specific key elements that are involved at the start of translation?

7. Draw the formation of a peptide bond.

8. Explain how polycistronic units coordinate expression of multiple genes.

9. Differentiate between an open reading frame and a coding sequence.

10. Using the codon table (Table 2.1) translate this DNA sequence into its corresponding amino acid sequence: 5′ - GGTGAAAATGAGACGATTTGTTCT - 3′. Using the single letter amino acid abbreviations, this should spell a word.

11. If the 14th base in the DNA sequence in question 10 became a T, how would this SNP alter the encoded protein sequence?

12. Compare designations of essential genes and accessory genes.

Discussion topics

These topics are presented for discussion in study groups, as part of class discussions, or on your own. These questions go beyond what is directly covered in this part of the book. Use the research literature and other reading to explore these topics in more depth. Tips to help prepare for topic discussions are available on-line.

13. There are many instances where bacterial genotypes do not match their phenotypes, where the physical features of the bacteria are different from what is encoded in its genome because one or more proteins are not being expressed. Discuss the reasons why a bacterial cell may not express particular proteins and what advantages this may have. Why would the genome continue to carry genes that are not being expressed?

14. One of the features that distinguishes bacteria is coupled transcription–translation. Explore how the linking of these two processes influences the expression and regulation of proteins within a bacterial cell, compared to what would occur in eukaryotic cells where these processes are physically separated.

15. Gene expression using alternative sigma factors can play a key role in virulence for many bacterial species. Discuss one species of interest where this is the case, considering the sigma factor, the genes it regulates, and the phenotypes this generates in the bacterial population.

Further reading

Genes are features in the DNA that encode proteins

Letunic I, Bork P. Interactive tree of life (iTOL) v3: An online tool for the display and annotation of phylogenetic and other trees. *Nucleic Acids Res.* 2016 Jul 8;44(W1):W242-5. doi: 10.1093/nar/gkw290. Epub 2016 Apr 19.

Snyder LA, Loman NJ, Fütterer K, Pallen MJ. Bacterial flagellar diversity and evolution: Seek simplicity and distrust it? *Trends Microbiol.* 2009 Jan; 17(1):1–5.

Waite RD, Struthers JK, Dowson CG. Spontaneous sequence duplication within an open reading frame of the pneumococcal type 3 capsule locus causes high-frequency phase variation. *Mol Microbiol.* 2001 Dec; 42(5): 1223–32.

Bacterial transcription generates RNA based on the DNA sequence

Iborra FJ, Jackson DA, Cook PR. Coupled transcription translation within nuclei of mammalian cells. *Science.* 2001; 293: 1139–42.

Laursen BS, Sørensen HP, Mortensen KK, Sperling-Petersen HU. Initiation of protein synthesis in bacteria. *Microbiol. Mol. Biol. Rev.* March 2005; 69(1): 101–23.

Bacterial translation produces proteins based on the mRNA sequence

Miller Jr OL, Hamkalo BA, Thomas Jr. CA. Visualization of bacterial genes in action. *Science.* 1970; 169: 392–5.

Bacterial mRNA can contain more than one gene and form operons

Lewis M. Allostery and the *lac* operon. *J Mol Biol.* 2013 Jul 10; 425(13): 2309–16.

Sigma factors are responsible for RNA polymerase promoter recognition

Feklístov A, Sharon BD, Darst SA, Gross CA. Bacterial sigma factors: A historical, structural, and genomic perspective. *Ann Rev Microbiol.* 2014; 68: 357–76.

Kazmierczak MJ, Wiedmann M, Boor KJ. Alternative sigma factors and their roles in bacterial virulence. *Microbiol Mol Biol Rev.* 2005 Dec; 69(4): 527–43.

Global regulators control the transcription of multiple genes

Perrenoud A, Sauer U. Impact of global transcriptional regulation by ArcA, ArcB, Cra, Crp, Cya, Fnr, and Mlc on glucose catabolism in *Escherichia coli. J. Bacteriol.* May 2005; 187(9): 3171–9.

Essential genes and accessory genes

Hutchison CA 3rd, Chuang RY, Noskov VN, Assad-Garcia N, Deerinck TJ, Ellisman MH *et al.* Design and synthesis of a minimal bacterial genome. *Science.* 2016 Mar 25; 351(6280): aad6253.

Genomes

This chapter will discuss the DNA within a bacterial cell that makes up the genome, including the contribution to the genome of genes that are essential to life. Genetic features within the bacterial genome, such as the rRNAs and tRNAs that are noncoding yet essential to life processes will also be discussed. This chapter will explore the concept of the genome in bacteria, where there is generally one circular chromosome that contains the whole genome, and how this is not true of all bacterial species. Some bacteria have multiple chromosomes that together make up the whole genome. Some bacteria have linear chromosomes as their genome or linear chromosomes that together with circular chromosomes make up the genome. Extra-chromosomal DNA can also be present within the bacterial cell, generally present as plasmids, which can contribute additional traits to the bacteria. The structure and composition of bacterial genomes will be explored in this chapter, including how these structures can contribute to the expression of genes and the consequences of transcription, translation, and replication happening simultaneously within the bacteria cell. Changes in the genome are also discussed, including how these can arise and therefore drive evolution of bacteria.

Bacterial chromosomes carry the genetic material of the organism

All of the DNA within a bacterial cell has the ability to contribute to the traits expressed by the organism. All genes within a cell have the potential to be expressed as proteins. This is the genetic material of the organism, the vast majority of which is within the bacterial chromosome, or chromosomes. The DNA present in the cell has to be replicated to continue on to the next generation of bacteria. The genetic features present in the DNA have to be transcribed to contribute to the traits expressed by the bacteria and the transcribed DNA has to be translated for the encoded proteins to influence the bacterial characteristics. Through replication, transcription, and translation the DNA is sustained in the species and can contribute to its characteristics.

When the genetic material within a cell is discussed, the term **genome** is used. A genome is generally considered to be all of the genetic material within a cell. In a bacterial cell, the genome may be contained within one molecule of DNA or several separate and discrete molecules of DNA. The genome is what makes up the **genotype** of the cell, being all of the genetic material in the cell encoding everything that the bacterial cell has the potential to express as its **phenotype**.

A **chromosome** is a discrete molecule of DNA, which can be in either a circular form or a linear form (**Figure 3.1**). In bacteria, chromosomes are most commonly found as circular DNA molecules, where the DNA double helix has no ends and forms a continuous circle. However, in some bacterial species linear chromosomes are present, where there are ends to the DNA double helix molecule in structures that are largely similar to what is seen in eukaryotes with their linear chromosomes.

Circular bacterial chromosomes may be depicted as circular maps, which show the relative location of genetic features around the circle of the DNA molecule

Figure 3.1 Bacterial chromosomes. Most bacterial species have a single circular chromosome (a). Some species have two or more circular chromosomes that make up the genome (b). A few bacterial species, however, have linear chromosomes that may constitute the whole of the genome (c) or be combined with circular chromosomes as the genome of these species (d).

Figure 3.2 A circular bacterial chromosome represented in different formats. Bacterial genome sequence data are represented in circular and linear formats. A circular display format reflects the circular nature of the bacterial chromosomal DNA (top). A linear display can often be easier to use for analysis, comparisons, and visualizations (bottom), where the circular DNA data has been linearized by selecting a first and last base. It must be remembered that the DNA is actually present as a closed circle in the cell, therefore features identified in analysis that are at the beginning (left end) are actually near to those at the end (right end) in the cell, where the chromosome is circular.

(**Figure 3.2**). However, it is also common to represent a circular chromosome as a linear molecule (see Figure 3.2). This is commonly done to make comparisons and alignments between circular chromosomes easier. It must be remembered in such cases that in actuality, the chromosome forms a circle, so the first base of the linearized form is actually next to the last base. In the past, it was common when reporting the sequence data to split a circular chromosome at the first base of the conserved gene *dnaA*, but this convention is not always observed. In cases where otherwise similar chromosomes are linearized starting at different places, the resulting alignments can look as if there has been a change in the genome, when this is really an artificial introduction of a break in the DNA sequence data and the presentation of a circular chromosome in an artificial linear form. Genome sequence data in the public databases will be linear in its format, regardless of the type of chromosome, with the first base chosen by the researchers designated as base 1.

Some bacteria have multiple chromosomes

The genome can be made up of more than one circular or linear chromosome or a combination of circular and linear chromosomes, depending on the bacterial species. For example, as mentioned in Chapter 1, *Brucella* and *Vibrio* species have two circular chromosomes and *Agrobacterium tumefaciens* has one linear chromosome and one circular chromosome. Regardless of the form of the chromosomes, in order to persist as the genetic material of the species, all of the chromosomes must be carried on to the next generation of bacterial cells. Each chromosome within the cell, being discrete and distinct portions of the genome, is replicated separately. As a result, each chromosome has its own origin of replication (discussed in Chapter 1), where the process of replication is initiated, and is copied by its own set of replication machinery.

Having multiple chromosomes that replicate independently, each with their own origin of replication, can mean that part of the genome is present in a higher copy number than the rest of the genome. One of the chromosomes may replicate more rapidly than the other, so that instead of there being one large circular chromosome and one small circular chromosome, the small chromosome is actually present in two or more copies, often due to the relative speed of copying each chromosome. In addition, with multiple origins of replication, more of the genome will be under the dosage effect of proximity to the origin. The concept of dosage effects and the origin of replication will be explored in more detail later in this chapter.

When multiple chromosomes are in a bacterial cell, whether circular or linear, the genome is divided between them. Part of the genetic material of the species is carried on one chromosome and part on another chromosome. This raises the question of what defines the DNA within a bacterial cell as a chromosome and therefore part of the genome. In order for the DNA molecule to be called a chromosome, it must carry essential genes. In some bacteria, we see all of the genes for certain essential functions, such as replication or transcription, on one chromosome while other essential genes are on another. The key factor in the chromosome definition being that there is at least one essential gene carried on that particular molecule of DNA.

Plasmids contribute additional genetic features

In the case of bacteria, it could be argued that **plasmids** are part of the genome because plasmids contain genetic material that is inside the cell. It would also be argued that plasmids are not part of the genome because plasmids are separate from chromosomes, can be acquired from other bacteria as a whole, and can be transient and therefore lost from a cell. Indeed, some plasmids can be gained or lost from a bacterial cell without impacting upon the organism. Other plasmids have evolved within the bacteria and carry genes that contribute to the characteristic traits that are attributed to the organism. Whether a plasmid should therefore be considered part of the genome or not is likely to be dependent on the bacteria in question and even then may be the subject of debate.

Plasmids contribute to evolution and diversity in bacteria through their presence. These circular DNA molecules bring to the bacterial cell alternative genes that are not present in the genome, which may open up new niches and enhance survival of the bacteria. In turn, the enhanced survival of the bacteria provides for the persistence of the plasmid, providing an environment in which the plasmid DNA is replicated and maintained. By carrying a trait that can increase survival of the bacteria, such as antibiotic resistance, the plasmid can not only ensure its continuation, but also its spread throughout the bacterial population. Such spread is then only limited by compatibility of the plasmid and its genes with the bacterial cell and its systems.

Additionally, some plasmids ensure their persistence through mechanisms such as the toxin/antitoxin systems, whereby a bacterial cell is destroyed if the plasmid is lost. These systems produce a long-lasting toxin within the cell that is neutralized by a short-lived antitoxin. So long as the genes for both are within the bacterial cell, it can survive. But, if the plasmid is lost, the action of the antitoxin stops before the action of the toxin and there is no genetic material from which to make more antitoxin. Therefore, the bacteria in the population that lose the plasmid are killed by the toxin in an effective mechanism that ensures that plasmid is kept in the bacterial population once it has been acquired. This is referred to as **post-segregational killing** (**PSK**) because it happens after the plasmid DNA has been segregated into the dividing bacterial cells.

When DNA that looks like a plasmid may actually be a chromosome

Some plasmids have been found that contain essential genes. Although plasmids that carry essential genes are likely to be much smaller in size than the chromosome(s) within the bacterial cell, it may be that the best designation for these sorts of plasmids is chromosome. The presence of essential genes differentiates a circular secondary chromosome from a **megaplasmid**, which is a very large plasmid. Like plasmids, chromosomes may have various numbers within cells. Chromosomes carry essential genes and genetic features, regardless of size. The designation of chromosome is therefore dependent on our ability to analyze the DNA sequence and identify essential genes. As the features of plasmids are explored in greater depth and the functions of the genes encoded are better defined, more essential genes may be identified. It is therefore possible that some DNA that was initially designated as a plasmid may be discovered to be better described as a small chromosome due to the presence of genes found to be essential.

Like a chromosome, each plasmid has its own origin of replication and goes through the process of replication independently of other DNA in the cell. Plasmid replication results in variation in **plasmid copy number**, or the number of copies of the plasmid that are present in the cell at any one time. Some plasmids may have just one copy in the bacterial cell, similar to the chromosome, while others may be present in 100–200 copies.

Prophages add bacteriophage genomes to a bacterial genome

Within the bacterial genome there may be one or more **prophages**, which are the genomes of bacteriophages, the small virus-like particles than can infect bacterial cells. During the bacteriophage lysogenic phase, its genome is integrated into the bacterial genome, where it is then carried, replicated, and passed on to new bacterial cells. The bacteriophage genome is thus able to make use of the biosynthetic systems within the bacterial cell to copy itself by becoming a prophage within the bacterial genome.

In addition to the bacteriophage genes, responsible for making new bacteriophage particles, the prophage genome may contribute genes to the bacteria that can influence the bacterial phenotype. These additional genes in the prophage genome contribute to the genetic material of the bacterial cell in a manner similar to the contribution of genes on a plasmid. Some bacteriophages are known to contain virulence factors and toxin genes that may alter the characteristics of the bacterial species once it acquires the prophage. Some of the prophages in bacterial genomes, with their additional genes, create strain-specific genotypes that are a differentiating factor between different strains of the same species that display different virulence characteristics.

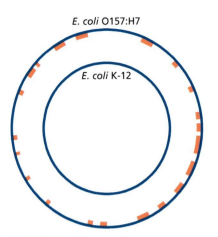

E. coli O157:H7

E. coli K-12

Figure 3.3 Differences in genome size between different strains of bacteria. Shown here is a schematic of the *E. coli* strain K-12 genome (inner circle) compared to the *E. coli* O157:H7 genome (outer circle). The regions in red are present only in the *E. coli* O157:H7 genome sequence and contribute to the difference in genome size between the bacteria of over 1 Mb.

Prophages can be major contributors to the content of a bacterial genome. In *Escherichia coli*, for example, the prophage genomes within the bacterial genome can account for up to 1 **Mb** (**megabase**) of the DNA in the cell, differentiating some strains of *E. coli* from other strains of *E. coli* with smaller genomes (**Figure 3.3**). There will be more on the contributions of bacteriophages to bacterial genomes later in the text, including an entire chapter dedicated to bacteriophages (see Chapter 21).

The sizes of bacterial genomes are characteristic of bacterial species

Each species of bacteria has a typical genome size and a typical number of genes, although not all genomes of the same species have the same genes. Within a bacterial species there are different **strains**, bacteria which vary from one another slightly on the genetic level, although they are all within the same species. The presence of prophage and other factors that may be strain-specific can alter the size of a genome beyond the scope of the typical genome size for the species. There will, however, be a typical size or size range for the genome of each bacterial species and comparisons against that size can reveal whether large changes in genome size, either insertions or deletions of genetic material, have occurred.

Regardless of the size of the genome itself, there is a general rule of thumb to approximate the number of genes within a bacterial genome. Due to the size of a typical bacterial gene and the length of DNA typically found between genes and associated with promoter regions, generally there is one bacterial gene per each **kilobase** (**kb**) of DNA. By this rule, it can be estimated that there are approximately 2,000 genes in a 2 Mb genome or 3,000 genes in a 3 Mb genome, which will be annotated as coding sequences (discussed in Chapter 2). Therefore, there is a correlation between the size of a bacterial genome and the number of genes that can be found within the genome (**Table 3.1**).

Contributions of the core genome to defining the species and the accessory genome to defining the strain

Although genome sizes can vary within a species, due to strain-specific differences in the presence of genes, prophages, and other elements, the core content of the genome will remain the same. This core content is a set of genes and other genetic material that is conserved among all bacteria within a species. This is the **core**

TABLE 3.1 Identified coding sequences and genome sequence size		
Bacteria	**Coding sequences**	**Genome size (bp)**
Sorangium cellulosum	9375	13033779
E. coli O157:H7	5449	5620522
Anaeromyxobacter dehalogenans	4346	5013479
E. coli K-12	4140	4641652
Bacillus subtilis	4175	4215606
Streptococcus pneumoniae	2125	2160842
Neisseria gonorrhoeae	2157	2153922
Haemophilus influenzae	1610	1830138
Helicobacter pylori	1469	1667867
Candidatus Zinderia insecticola	206	208564
Candidatus Nasuia deltocephalinicola	137	112091

genome (**Figure 3.4**). Specifically, the genes that are conserved within a bacterial species are the **core genes** whereas the core genome includes other conserved genomic content, such as rRNAs and tRNAs.

It is the genes within the core genome that characterize and define the bacterial species, as modified and regulated by the other noncoding genetic features. This core genetic material will encode the traits that are typically expressed by all of the strains within a species and determine how and when these traits are typically expressed. The core genes tend to be fundamental to the bacterial species, including those that are essential for life and those that are important for survival in the typical niche occupied by the bacterial species.

Different strains within the same species may also possess other genes and other genetic material that may not be shared across the whole of the species. This is referred to as the **accessory genome** because it contains other genes and features that are not present in all representatives of the species (see Figure 3.4). These genes, considered **accessory genes**, can include virulence factors that vary between strains, producing particular virulence phenotypes such as the production of strain-specific toxins. The accessory genome can also include genes or regulatory systems that enable some strains within a species to survive in different environments than the standard niche for the species or features that enable some strains to withstand environmental changes in a niche that are not characteristics of the whole of the species.

All of the possible genes within a single bacterial species is the **pan-genome** (see Figure 3.4). It includes all of the core genome genes and all of the accessory genome genes that have been identified. Some accessory genome genes may never actually be found all together in the same bacterial cell and may indeed be mutually exclusive, where their presence in the same genome is not compatible. However, the species as a whole has the potential to carry any of the accessory genome genes along with the core genome genes, therefore they are all possible genes of the species and included in the pan-genome.

Bacterial genomes are densely packed

Bacterial chromosomes are fairly densely packed with genes and other genetic features, including rRNA loci, small noncoding RNAs, and prophage. Compared to eukaryotic genomes, there is often little space, very few base pairs, between each of these features. Regions previously believed to be longer stretches of intergenic DNA, being between two genes, are being revealed to contain small noncoding RNAs and/or important regulatory sequences. As such, the bacterial genome is very densely packed, especially when all potential genetic features are considered. The distribution of features in a bacterial chromosome tends to be fairly even across the genome. This means that the architecture of the genome, including its content, homology, and synteny in comparison with other sequences, and the architecture of the individual chromosomes can be analyzed easily, as there will be very few gaps, and contribute to understanding the nature of the genome and its contents.

DNA base composition differs between species

Most of the chromosome, as a whole, will share the same composition with regards to the proportion of each of the nucleotide bases that it contains, the number of adenines and thymines and of guanines and cytosines. Erwin Chargaff noted in 1950, that DNA was made of the bases adenine, cytosine, guanine, and thymine, but that these bases were not equally represented in all bacterial species. The variations in the proportions of adenines and thymines and of guanines and cytosines can be illuminating.

The number of guanines and cytosines (**G+C**) in bacterial genomes ranges from 14% in *Candidatus Zinderia insecticola* to 75% in *Anaeromyxobacter dehalogenans* of the total genomic composition. This means that most of the bases in *Z. insecticola* are adenine and thymine base pairs and that most of the bases in *A. dehalogenans* are guanine and cytosine base pairs. The characteristic G+C can give us insight into the bacterial species and its behaviors. In general, the G+C within the genomes of

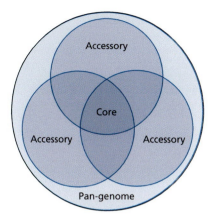

Figure 3.4 The core genome, accessory genome, and pan-genome. The core genome includes all of the genetic features that are included in all of the examples of a particular species. Each genome will, in addition, have other genetic features that contribute to the accessory genome. When all of the potential genes in a species are considered, everything from the core genome and all accessory genomes, even when not seen together in the same cell, this is the pan-genome.

free-living bacteria is higher than that found in bacteria that cannot survive outside their hosts. Therefore, the G+C calculation can reveal something about the typical growth environment of the organism being studied.

The cost of synthesizing the raw materials required for guanines and cytosines, the GTP and CTP within the cell, is higher than the cost of synthesizing the ATP and TTP for adenines and thymines. This cost difference works against genomes having a high G+C content from an energy point of view. As a result, bacterial species that inhabit nutrient-limited niches where energy may be a limiting factor, tend toward lower G+C content.

In addition, larger genomes, such as those greater than 3 Mb, tend to show higher G+C than smaller genomes. Therefore, the genome size should be taken into account when assessing the G+C and its likely implications, as this could be a reflection of size rather than just habitat.

Base composition differs between coding region

The G+C is not universal across the whole of the genome. For example, the conserved function of rRNA means that DNA sequences of these loci tend to have slightly different G+C to the rest of the genome. Coding sequences and intergenic regions, however, will tend to adhere to the overall G+C of the genome. Alteration of a gene sequence to suit the genomic G+C average is far less likely to have a detrimental impact upon the organism than would changing the sequence of the highly conserved rRNA. The availability of different codons corresponding to the same amino acid means that codons within gene sequences can be changed to synonymous codons without changing the amino acid sequence encoded by the gene, therefore providing scope for changing the gene to suit the G+C of the genome.

The origin of replication impacts the base composition

The **origin of replication** is the location along a chromosome where replication begins. It is a region that is recognized by the replication machinery and where replication will consistently start. This impacts upon the nature of the chromosome, including what genes are present and what is encoded in this region, the orientation of genes in this region, and the base composition of the DNA in this region.

Near the bacterial chromosome's origin of replication, the DNA sequence tends to be higher in guanines and cytosines, while there tend to be more adenines and thymines near the **terminus**, where replication ends. In addition, there is a bias for bases depending upon if they are on the **leading strand** and **lagging strand** of replication and it is this bias that can help identify the origin of replication.

There are differences in the mutations that can accumulate on the leading and lagging strands of replication, introduced due to the differences in replication mechanisms on each strand. As a result, the leading strand tends to have more guanines, while the lagging strand tends to have more cytosines. Through recognition of these leading and lagging strand biases, it is possible to identify the origin of replication on the bacterial chromosome and the **termination of replication**, the location where a round of chromosomal replication ends.

The calculation of G−C/G+C for the leading and the lagging strand is the **GC skew**. Trends in the GC skew can identify the origin of replication. As seen in Chapter 1, the leading and lagging strands at the two replication forks are on opposite strands, resulting in the GC skew. Assessing the GC skew around a circular chromosome can create a recognizable pattern that assists in the definition of the origin of replication for the chromosome (**Figure 3.5**).

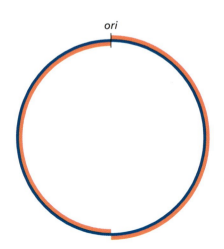

ori

Figure 3.5 The GC skew of a circular bacterial chromosome can identify the origin of replication. The calculated G−C/G+C (red) is here mapped to the circular chromosome (blue). Mapping of the G−C/G+C data onto the circular chromosome reveals the bias in base composition for the replication leading (right) and lagging (left) strands. Therefore, the origin of replication can be inferred from this data (ori).

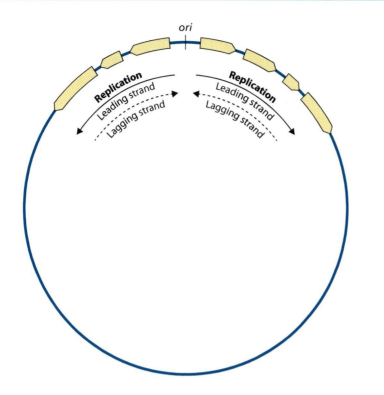

ori

Figure 3.6 Presence of genes on the leading strand and the lagging strand. The origin of replication for this circular genome is shown and the direction for the leading strand of replication indicated. The genes and their orientation on the chromosome are illustrated, showing that most of the genes are on the leading strand, being encoded divergently from the origin of replication. These leading strand genes are transcribed in an orientation that goes away from the origin of replication.

The influence of the leading strand and the lagging strand on gene orientation

There tend to be more genes on the leading strand of the bacterial chromosome versus the lagging strand, due to the continuous nature of replication on the leading strand. Although genes can be encoded on either strand, there is a bias toward genes being present on the leading strand, which is replicated continuously. The presence of a greater number of genes on the leading strand creates a strand bias for the orientation of genes in the chromosome, which can be observed when analyzing the genome sequence (**Figure 3.6**).

This abundance of genes on the leading strand is the result of evolutionary pressures. The observed strand bias against genes on the lagging strand is likely to be due to the progression of both DNA replication proteins and RNA transcription proteins on the same DNA strand. Inside the bacterial cell the processes of transcription and replication are happening at the same time, using the same chromosome as their templates for copying the DNA and for generating mRNA from the DNA. The processes progress at different rates along the DNA and as a result they are likely to collide as they are conducting replication and transcription, respectively.

When replication and transcription are both going in the same direction, on the leading strand, there is a chance of collision between the proteins involved when they catch up to each other. However, collisions are perhaps either more frequent or more disruptive on the lagging strand because of the discontinuous nature of replication on the lagging strand, regardless of the speed of the processes. All collisions between replication and transcription on either strand will slow the progress of the replication fork.

Highly transcribed genes are likely to be being transcribed during replication. The transcription of highly expressed genes will therefore cause fewer collisions if the genes are located on the leading strand, rather than the lagging strand. Indeed, the most heavily transcribed regions of the genome are the rRNA loci, which are present on the leading strand. To avoid or minimize the effects of clashes between transcription mechanisms and replication mechanisms, most genes are also encoded on the leading strand.

Essential genes tend to be on the leading strand, regardless of their expression level. Mutations can occur as a result of replication stalls following a polymerase collision. Perhaps the positioning of essential genes on the leading strand is beneficial

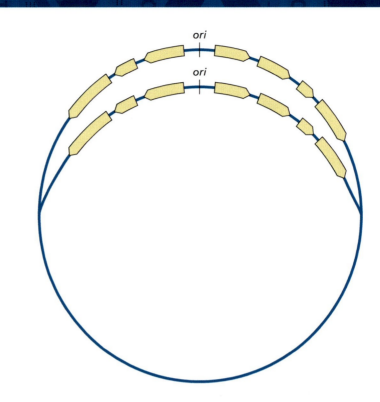

Figure 3.7 The dosage effect upon genes of their presence near the origin of replication. When genes are present close to the origin of replication (ori), this can impact the number of copies of the gene present in the bacterial cell. As replication progresses, the part of the bacterial chromosome near the origin of replication is present in more than one copy. Thus, the genes in this region are present in the cell in more than one copy, each capable of being a template for transcription. When there is more than one chromosome, gene dosage can affect more genes due to the availability of origins of replication on each chromosome.

because this placement is more likely to avoid the introduction of mutations into essential genes during such replication stalls.

Genomic architecture can impact gene expression

The location of a gene within the chromosome can impact upon its expression. For example, during replication, genes that are near to the origin of replication are copied first. This creates two copies of these genes in the cell, increasing the gene dose. Replication does not necessarily wait for one round to end before another begins, so as further rounds of replication initiate, more copies of the genes near the origin are generated. The genes near the origin therefore experience a **dosage effect** and tend to be the most highly expressed genes in the genome (**Figure 3.7**).

Although the process of replication is about copying the genetic material within the cell, it can also impact upon the process of transcription, which generates mRNA. On the lagging strand, the progression of DNA polymerase during replication displaces any RNA polymerase that is in its way. This stops all transcription by the RNA polymerase from the genomic region until replication moves on. Therefore, the expression of genes on the lagging strand is prone to disruption. The impact on the leading strand is less disruptive.

Conservation of the order of genetic features between bacterial species

Homology is a measure of the similarity between sequences at the DNA or protein level. Another term, **synteny**, is used to describe similarity in the presence and order of features within a chromosome. Synteny refers more to the structure and content of regions than it does to their actual sequence.

Early in genomic investigations, it became apparent that genes encoding proteins involved in associated functions tended to cluster together on the chromosome. The close proximity within the chromosome of genes that contribute to related functions appears to be beneficial for the co-regulation of these genes. Adjacent genes may be co-transcribed into the same mRNA. These co-transcriptional units can form operons when they are co-regulated. There will be more on operons in Chapter 5. When analyzing a genome, it is important to consider whether adjacent genes may be co-transcribed, as this can

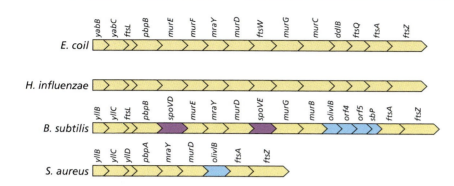

Figure 3.8 The chromosomal synteny of the *dcw* **cluster between diverse species.** Shown here are the division and cell wall synthesis clusters of genes for *E. coli*, *H. influenzae*, *B. subtilis*, and *S. aureus*. These species are quite diverse, therefore, the homology between the genes of the different species can be quite low. However, the overall organization and composition of the *dcw* clusters are similar, making them a region of synteny between these species.

reveal how the genes are regulated. Evolutionary pressures will tend to keep these blocks of co-transcribed and co-regulated genes together. For example, if a block of genes that function together are separated by a rearrangement of the order of genetic features in the chromosome, there may be a loss of fitness for the affected bacteria. The cells affected by the rearrangement will be selected against in the bacterial population and those that retained the original order of the genes will be selected for in the population. Therefore, evolutionarily, regions that contain clusters of genes with related functions are preserved and this synteny can be seen across even very diverse bacteria.

One striking example of synteny that is retained in many bacteria, even those that are evolutionarily very distant, comes from common features of bacterial cells. All bacterial cells must divide into daughter cells to grow and survive. Most bacterial cells have peptidoglycan cell walls. Construction of more of the peptidoglycan cell wall is a coordinated event that coincides with the process of cellular division; as the cell grows to enable it to then divide, a greater quantity of cell wall material is required. Therefore, in most bacterial species the genes involved in cellular division and those involved in cell wall synthesis are clustered together. Chromosomal synteny can be seen at the division and cell wall synthesis (*dcw*) locus. There is similarity in the genes and their organization across species barriers at the *dcw* cluster, even when very diverse bacteria that are evolutionarily distant are considered (**Figure 3.8**). The *dcw* clusters therefore have synteny, because the genes and their configuration are conserved between bacterial species, even if the sequence homology calculations show that the sequences are not very similar. Other gene clusters that remain syntenic across evolutionarily diverse species include the NADH dehydrogenase operon (*nuo* genes) and the ribosomal protein operons.

Distantly related species may only have a few key gene clusters that are conserved. Comparisons of closely related bacterial species will reveal greater synteny between the chromosomes. When assessing a region of a chromosome and comparing two or more species, synteny is one of the features investigated. If two species share synteny over a region, then they may share in the function of the genes encoded in the syntenic regions. Comparing genomes, there may also be a gene that is present in just one strain or species in a region that otherwise has synteny; this may reveal something about differences between the bacteria. Changes in synteny, in the order in which genes are found in a chromosome, may indicate differences in the regulation and expression of genes between two strains or species.

Supercoiling can also influence gene expression

The DNA that makes up the genome and any additional genetic material, such as plasmids, must all be contained within the bacterial cell. Each 10 bases in a DNA double helix is 3.4 nm long. Bacterial genomes can range in size from 112,091 bases (in *Candidatus Nasuia deltocephalinicola*) to 14,782,125 bases (in *Sorangium cellulosum*), therefore, the genomes can range in size from approximately 38,111 to 5,025,923 nm. Bacterial cells range from 200 nm (in *Mycoplasma gallicepticum*) to 750,000 nm (in *Thiomargarita namibiensis*). The DNA that must be inside the cell is often far larger than the cell itself. In order to fit all of the genetic material inside of the bacterial cell, the DNA must be compacted. To achieve this, bacterial cells will

Genome = 1,578 µm

Bacteria =   0.5 µm
            2 µm

Figure 3.9 Supercoiling of DNA compacts its size. Shown is the space taken up by a DNA molecule of 1 Mb if it is stretched out straight. Following supercoiling, the same 1 Mb of DNA takes up far less space and is able to fit inside the bacterial cell.

Nucleoid

Figure 3.10 The mass of supercoiled DNA that is the bacterial nucleoid. Supercoiling of DNA compacts the genome so that it fits inside the cell, producing a mass of condensed genetic material in the cell that can be seen and is identified as the nucleoid.

supercoil their DNA to make a very long molecule into a mass of DNA that is far more compact. There are some proteins that aid in the process of supercoiling, help retain supercoiling, or otherwise aid in bending and compacting the genome, which will be introduced in Chapter 5. To achieve a compact state and fit within the bacterial cell, most of the DNA is negatively supercoiled, twisted against the normal turn of the helix such that it becomes much smaller and able to fit within the cell (**Figure 3.9**).

Unlike eukaryotic cells, bacterial cellular DNA is not contained within the membranes of a nucleus. Bacteria do not have membrane-bound organelles such as nuclei. However, due to the supercoiling of the DNA into a compact mass of genetic material, a **nucleoid** can be observed. Although the nucleoid is not a discrete organelle, the nucleoid is a location within the cell that contains the DNA (**Figure 3.10**).

The compact structure of the bacterial chromosomes can limit the availability of the sequences of some genes for transcription, depending upon where the genes are located within the mass of DNA. It has been observed in experiments with *E. coli* that when the cell is treated with a chemical to halt transcription and translation, the nucleoid becomes a very compact sphere. However, the processes of transcription and translation need access to the DNA strands, which become more loosely associated with the nucleoid, associating instead with transcription and translation machinery. This gives the nucleoid its irregular appearance in the cell. Supercoiling of the DNA can make it difficult for RNA polymerase to access promoter sites and initiate transcription for regions of the chromosome buried deep within the compacted nucleoid structure. Likewise, genes that are highly expressed can impose limits on the compaction of a genomic region, due to the continual presence of the coupled transcription–translation machinery.

Distribution of noncoding genetic features in the bacterial chromosome

In addition to genes, there are noncoding genetic features present within the chromosomes of bacteria, including the locations that encode the tRNAs. These features are carried within the DNA and transcribed to the tRNAs that carry amino acids to the ribosome. There can be multiple tRNA loci in different locations in the genome for each amino acid, with each tRNA carrying a unique anticodon that corresponds to the different codon possibilities for each amino acid.

The rRNAs that are present within the ribosomes are also carried in the DNA genome. These loci contain the sequences for all of the ribosomal RNA. The rRNA locus is transcribed and the rRNAs are incorporated into the bacterial 70S ribosome along with ribosomal proteins. Together these components make up the 50S ribosome, with its 23S and 5S rRNA, and the 30S ribosome, with its 16S rRNA. In bacteria, there can be several copies of the rRNA locus present around the genome (**Figure 3.11**), each carrying the sequences for 23S, 5S, and 16S rRNAs. These can be scattered around the single circular chromosome, creating regions within the chromosome of DNA sequence identity (or near identity).

Further, there are small RNAs present as genetic material within bacterial cells that are transcribed but not translated. These **noncoding RNAs (ncRNAs)** tend to be present in intergenic regions or as regions overlapping coding regions, but on the opposite strand. In some cases, ncRNAs have been identified as genes in error, being labeled in the genome sequence data as coding regions. Functional studies and increased understanding and recognition of ncRNAs can help clarify the assumptions made about the nature of features in genome sequence data. Regions previously believed to be "junk DNA," or devoid of genes, have been discovered to contain ncRNAs that serve vital regulatory functions within the cell. Further study of these and other yet to be discovered features of bacterial genomes may reveal even more within the sequence of DNA.

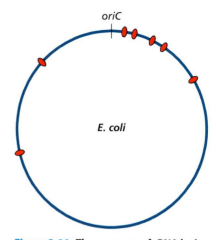

Figure 3.11 The presence of rRNA loci in bacterial genomes. Some bacterial genomes, like the *E. coli* shown here, have more than one rRNA locus in the genome. Here there are seven loci in this single circular chromosome, with each locus encoding the 5S rRNA, 16S rRNA, and 23S rRNA (indicated in red).

Mutations in the bacterial genome

The bacterial genome is subject to mutations, changes in the genetic code that may, or may not, alter the encoded features. **Mutation** drives bacterial evolution. Through the changes to the organism's genetics, bacteria are able to expand into new niches and display new characteristics. A change to the genome as a result of mutation can increase, decrease, or have no effect on the fitness of the organism.

Transposons, IS elements, and other mobile elements contribute to genomic changes

There are other genetic elements in the DNA that contribute to changes within a bacterial genome, many that differ in presence or location between strains of the same bacterial species. **Transposons**, for example, are mobile genetic elements capable of self-excision and reintegration into the chromosome. Transposons carry within their sequences all of the genes necessary for mobilization to new parts of the bacterial genome. Transposons can also carry other genes, such as antibiotic resistance determinants.

Insertion sequence (IS) elements are related to transposons in structure and mobilization, however, **IS elements** do not carry additional genes. Like transposons, IS elements can move from one part of a bacterial genome to another. Introduction of new genetic material at a site in the genome through the insertion of an IS element can disrupt the genes of the region or their regulation. Both transposons and IS elements can contribute to genetic mutation by disrupting genes in this way.

Other mobile elements include **prophage**, which are capable of self-insertion during the bacteriophage lysogenic phase and which excise from the chromosome during the lytic phase of bacteriophage growth. Prophage insertion into the genome may disrupt genes and regulatory networks in a manner similar to the insertion of transposons and IS elements. Like transposons, the introduction of a prophage genome may also bring with it new genes.

Small changes in the genome created by point mutations and indels

The process of replication is not perfect. As mentioned, there can be stalls in replication. These events may cause errors to be introduced into the DNA sequence. In addition, the replication machinery itself can sometimes introduce errors. While there are error correction systems within the bacteria to mitigate against the introduction of replication errors, not all errors are corrected. Therefore, changes to the sequence of the genome can be introduced during the process of replication. In addition, some bases can change spontaneously to another base. Unless it is

Figure 3.12 The deamination of cytosine to uracil generates a point mutation. Cytosine, by the addition of H_2O, loses NH_3 and becomes uracil. Uracil hydrogen bonds with adenine, creating a point mutation.

corrected, the spontaneous change in a base can become fixed into the DNA sequence. These events create point mutations, changes to single base pairs along the double helix.

The most common **point mutation** is cytosine to thymine. This change occurs as a result of the spontaneous **deamination** of the cytosine, forming uracil (**Figure 3.12**). Uracil is not a base that is normally found in DNA, so during replication the uracil is copied as a thymine and thus there is a change from a cytosine to a thymine. These are not the only point mutations that occur. Any base can erroneously be changed, some as uncorrected errors in replication, some as chemical changes to bases, and some as errors introduced by repair mechanisms.

By the same mechanisms that introduce point mutations, another type of mutation called indels can occur. An **indel** is a short insertion or deletion of one or a few bases. These small changes can alter the reading frame of a coding sequence, frame-shifting a gene so that it prematurely reaches a termination codon. They can also disrupt regulatory regions, altering the binding sites for RNA polymerase and/or regulatory proteins.

Often when comparing genome sequences these changes can be observed, both single base changes in sequences and small insertions or deletions. Since it can be difficult when analyzing genome sequence data to correctly identify the original sequence, to determine if an insertion or a deletion has occurred, the term indel covers both possibilities. Likewise, a single base change is called a **single-nucleotide polymorphism** (**SNP**). The designation SNP is not necessarily a judgement on which base is mutated when comparing two genomes, it is simply an indication that the bases are different.

Translocations can change the order of genetic features in a genome

Translocation is movement of DNA from one place to another along or between DNA molecules. The transfer of DNA from one location on a DNA molecule to another location on the same molecule, from one chromosome to another, from chromosome to plasmid, or from plasmid to chromosome, are all examples of translocation.

The key feature of translocations is the transposition within a chromosome or plasmid, where a segment of DNA is moved from one location to another. Translocation does not alter the genetic content of the cell, although genes may be disrupted in the process of the DNA moving if the borders of the translocation are within genes.

A change in the position of a gene within the chromosome can alter the expression of the gene through a variety of mechanisms. Altering the proximity of the gene from the origin of replication may either increase or decrease the gene dose, therefore providing more or fewer templates for transcription. A change in the strand upon which the gene is located can also impact upon its expression, with genes on the lagging strand being more prone to interference during transcription by the procession of the replication machinery. Translocation may also modify the regulation of the expression of a gene through causing a change in the relative location of genes from their regulatory elements. It may be that a promoter was disrupted by the translocation or that there has been an increase in the relative distance a regulatory protein would need to diffuse in the cell from where it is transcribed and translated to its binding site of its regulon.

Figure 3.13 Inversion of DNA changes the strand of DNA upon which a sequence is located. Three genes are shown here, all encoded on the top strand. An inversion occurs, represented by the parentheses. The middle gene is now in the opposite orientation and on the bottom strand.

Inversions flip the DNA strand upon which genetic features are located

Inversion is a form of translocation, where a DNA region is inverted, resulting in the DNA carried in the chromosome or the plasmid being in the opposite orientation of how it was originally. In this type of translocation, the DNA remains on the same DNA molecule, but is reintegrated into it so that the strands of the DNA are switched. What was on one of the strands will then be on the opposite strand (see Figure 3.12). Inversions can occur anywhere in the chromosome, however, due to the consequences of the inversion, not all remain fixed. Some inversions may reduce the fitness for survival of the bacteria and therefore are selected against evolutionarily.

Inversions that occur around the origin of replication can preserve both the synteny of the regions involved in the inversion and also preserve any gene dose effect produced by the proximity of the genes to the origin. In addition, the location of genes on either the leading or lagging strands is also preserved when an inversion is centered around the origin of replication (**Figure 3.13**).

Inversions can be programmed events involved in gene-expression changes. In these cases, the bacterial cell possesses inversion mechanisms that are capable of recognizing the invertible region and completing its inversion. Generally, inversions can result in genes being placed and removed from the influence of a promoter region (**Figure 3.14**).

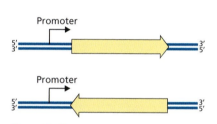

Figure 3.14 Programmed inversions can influence gene expression. In some bacteria, inversion events may place a gene, such as shown here, under the control of a promoter (bent arrow) in one orientation (top) or not in the inverted orientation (bottom). Therefore, the gene is only expressed in one configuration, subject to the inversion event.

Recombination changes the genome

Recombination is a term to describe the exchange of genetic material between or within DNA molecules. This process makes possible rearrangements, deletions, inversions, and other changes to the DNA. Crucial to the process is the joining together of DNA strands that were not previously joined. In bacteria, this can occur within a chromosome, between chromosomes when the bacterial cell has more than one, between copies of the chromosome during replication, between chromosomes and plasmids, and between the bacterial DNA and fragments of DNA that have entered the cell.

Generally, **homology**, the similarity between sequences, can facilitate recombination. The homologous regions, where the DNA is the same or similar, will come together and be acted upon by the recombination machinery, resulting in a new DNA sequence. Non-homologous recombination can also occur, bringing in DNA sequence to a region with which it does not share sequence similarity.

Through recombination, the DNA within bacterial cells can change. Although these organisms do not reproduce sexually, they are capable of mixing, shuffling, and rearranging genetic information in their chromosomes (**Figure 3.15a**), which gives them scope for evolution and adaptation. In addition to the genetic material within the bacterial cell, recombination can act on DNA that has entered the cell from elsewhere, further enhancing the potential for change to the genome (**Figure 3.15b**).

Horizontal gene transfer introduces new genetic material

Most of the changes to a bacterial genome can be attributed to **horizontal gene transfer** (**HGT**). This is a process whereby DNA from another organism is taken

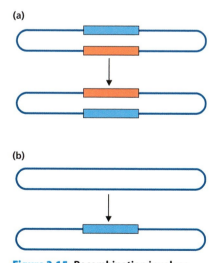

Figure 3.15 Recombination involves the exchange of DNA between two molecules. This can rearrange the sequence in the chromosome (a) or introduce new genetic material (b).

Figure 3.18 Bacteriophage can transfer DNA from one bacterial cell to another in a process called transduction. A bacteriophage attaches to a bacterial cell and transfers into the cell its bacteriophage genome (left images). The bacteriophage genome directs the bacterial cell to make more bacteriophage particles (middle image). These particles are packaged with bacteriophage DNA but may also package bacterial DNA. When these bacteriophages infect a new bacterial cell, they carry the bacterial genetic material to the new cell (right images).

To be a donor for conjugation, the bacteria must possess certain genetic traits, referred to as the **fertility factor** or **F-factor**. Generally, this is contained on an **F plasmid**, which through the conjugation machinery is able to transfer a copy of itself from the donor bacterial cell to the recipient cell, therefore establishing the F plasmid in the recipient. The recipient then becomes a donor, a cell capable of conjugation.

Recombination can, on occasion, bring the F plasmid into the chromosome of the bacterial cell. These are referred to as **Hfr** strains for **high frequency of recombination**. When an Hfr strain engages in conjugation, it takes with it the DNA into which it has integrated, therefore transferring chromosomal DNA from one cell to another, rather than plasmid DNA. The transferred chromosomal DNA, with its homology to the chromosome within the cell, has a high frequency of recombination with the recipient chromosome.

The process of conjugation requires direct contact between the bacterial cells and therefore can be disrupted if the connection between the cells through the conjugation pilus is lost. This phenomenon has been exploited in mapping experiments to identify the location of different genetic features on a chromosome. When located in this way, the feature is said to be at a certain number of minutes on the chromosome, corresponding to the time it takes for a feature on an Hfr chromosome to transfer to a recipient cell.

The process of transduction can introduce bacteriophage DNA into a bacterial cell

Transduction can transfer DNA from one bacterial cell to another via bacteriophages. Often the transferred DNA is bacteriophage DNA, however, it is possible for bacterial genomic DNA to be transferred in this way (**Figure 3.18**).

During the process of making new bacteriophage particles within the bacterial cell, the bacteriophages are assembled and filled with DNA. This process may result in some bacterial DNA being packaged into a bacteriophage particle. In this way, the bacterial DNA can be moved from one bacterial cell to another. This process, unlike conjugation, does not require the bacteria to be in contact with one another, as the bacteriophage acts as the intermediary in the process of moving the DNA from cell to cell.

Bacteriophages dock with bacteria through interaction with the bacterial surface, therefore, these transfers are limited by bacteriophage specificity. Transfer of DNA from one bacterial cell of a species to another of the same species is more likely to occur in transduction than transfer to unrelated species.

Key points

- Bacteria can have one or more chromosomes that make up the genome, containing the genetic material of the cell.

- The size of a bacterial genome can be influenced by its environment, yet the size is generally characteristic of the species.

- Bacteria can have extra-chromosomal DNA, such as plasmids, that carry accessory genes.

- There is roughly one gene for each 1 kb of DNA sequence.

- The genes conserved in a species and found in all examples of the species make up the core genome, with other genes making up the accessory genome.

- The size of the genome and the environment in which the bacteria are generally found will influence the number of guanines and cytosines found in the DNA sequence.

- The origin of replication will influence the expression of genes through gene dose and through the orientation of genes on the leading and lagging strand.

- The introduction of transposons, IS elements, and prophage into regions of the genome may incorporate new genes into the bacterial cell and may disrupt existing genetic features.

- Small changes in the DNA sequences between closely related bacteria, such as SNPs and indels, are likely to have occurred due to replication errors and other mutagenic events.

- Horizontal gene transfer can result in the incorporation of new DNA into the genome, resulting in regions of difference between strains of the same species.

- Transformation, conjugation, and transduction are naturally occurring mechanisms for bacterial uptake of DNA from outside sources.

Terms, questions, and discussions

Key terms

Define the following terms introduced in this chapter. Check your answers using the definitions in the Glossary. These terms are also available as Flashcards on-line.

Accessory genes	Hfr	Point mutation
Accessory genome	HGT	Recombination
Conjugation	Indel	Regions of difference
Conjugation pili	Inversion	Strain
Core genes	IS elements	Supercoil
Core genome	Kilobase	Synteny
Deamination	Megaplasmid	Termination of replication
Dosage effect	Mutation	Transduction
F plasmid	Nucleoid	Transformation
G+C	Pan-genome	Translocation
GC skew	Plasmid copy number	Transposon

Self-study questions

Answer each question using 50–100 words or a table or labeled diagram. Advice on where to find answers to these questions is available on-line.

1. What is the term for a bacteriophage genome that has been incorporated into a bacterial genome and what types of genes might be found there?

2. What are the typical size ranges of bacterial genomes and what features can alter genome size within related bacterial genomes?

3. How is it possible that the entire length of the bacterial genome is able to fit inside of the bacterial cell?

4. In general, what is the most likely orientation of genes near the origin of replication?

5. What is the term used to describe the similarity between the presence and order of features in the genome and how does this compare to homology?

6. What do we now know is present in regions once believed to be "junk" DNA?

7. Two genetic elements are able to mobilize in bacterial genomes. What are these elements and how does their movement introduce mutations?

8. Draw the structures involved in the deamination of cysteine to uracil.

9. Compare and contrast the difference between an indel and a SNP.

10. What are the consequences of translocation of DNA in a bacterial genome?

11. Draw a diagram of how an inversion event can influence the expression of a bacterial gene.

12. Bacterial cells can acquire new DNA via three mechanisms. What are they and briefly how do they work? Which one requires direct contact between bacterial cells?

Discussion topics

These topics are presented for discussion in study groups, as part of class discussions, or on your own. These questions go beyond what is directly covered in this part of the book. Use the research literature and other reading to explore these topics in more depth. Tips to help prepare for topic discussions are available on-line.

13. In Chapter 2, the concepts of essential genes and accessory genes were introduced. In this chapter, the concepts of core genes and the core genome were added to these terms. Compare the differences in meaning between the terms essential gene and core gene. Discuss the types of genes that might fall into both categories or only one.

14. The term pan-genome is used to describe all of the known genes of a particular species of bacteria, or sometimes more than one related species. As we sequence more genomes, are pan-genomes getting larger? Do you think we will achieve a definitive pan-genome for species that participate in horizontal gene transfer?

15. Mutation drives evolution. In bacterial species, there are many different mechanisms that can introduce mutations in the genome. Evaluate the various types of mutations and how they have contributed to bacterial evolution of key phenotypes, such as virulence or antibiotic resistance.

Further reading

DNA within the bacterial cell

Canchaya C, Fournous G, Brüssow H. The impact of prophages on bacterial chromosomes. *Mol Microbiol*. 2004 Jul; 53(1): 9–18.

Hayashi T, Makino K, Ohnishi M, Kurokawa K, Ishii K, Yokoyama K *et al*. Complete genome sequence of enterohemorrhagic *Escherichia coli* O157:H7 and genomic comparison with a laboratory strain K-12. *DNA Res*. 2001 Feb 28; 8(1): 11–22.

Genome architecture

Database of Genome Size www.cbs.dtu.dk/databases/DOGS.

Land M, Hauser L, Jun S-R, Nookaew I, Leuze MR, Ahn T-H *et al*. Insights from 20 years of bacterial genome sequencing. *Funct Integr Genomics*. 2015; 15(2): 141–161.

Pomerantz RT, O'Donnell M. What happens when replication and transcription complexes collide? *Cell Cycle*. 2010; 9(13): 2537–43.

Rocha EPC. The organization of the bacterial genome. *Annu Rev Genet*. 2008; 42: 211–33.

Supercoiling

Griswold A. Genome packaging in prokaryotes: The circular chromosome of *E. coli*. *Nat Educ*. 2008; 1(1): 57.

Mutations in the bacterial genome

Block DH, Hussein R, Liang LW, Lim HN. Regulatory consequences of gene translocation in bacteria. *Nucleic Acids Res*. 2012 Oct; 40(18): 8979–92.

Horizontal gene transfer introduces new genetic material

Barlow M. What antimicrobial resistance has taught us about horizontal gene transfer. *Methods Mol Biol*. 2009; 532: 397–411.

Finkel SE, Kolter R. DNA as a nutrient: Novel role for bacterial competence gene homologs. *J Bacteriol*. 183(21): 6288-93.

Huddleston JR. Horizontal gene transfer in the human gastrointestinal tract: Potential spread of antibiotic resistance genes. *Infect Drug Resist*. 2014 Jun 20; 7: 167–76.

Transformation involves bacterial uptake of DNA from its surroundings

Blokesch M. Natural competence for transformation. *Curr Biol*. 2016 Dec 5; 26(23): 3255.

Part II

RNA, Transcriptional Regulation, and Transcriptomes

The following three chapters will explore in more detail the role of RNA in the bacterial cell, the regulation of transcription of various RNAs in the bacterial cell, and the study of the whole of the transcripts within the bacterial cell that make up the **transcriptome**. These chapters will build on material presented in Part I of this book, presenting the material introduced there in more depth and detail.

RNA

RNA, ribonucleic acid, is a key biological molecule and in bacteria it is responsible for a variety of functions within the cell. The RNA from bacterial cells has been investigated using a variety of techniques, starting from the simple isolation of the nucleic acids away from the other components of the cell to understanding the sequence order of the ribonucleotides. Each of the types of RNA (mRNA, tRNA, rRNA, and ncRNA) has differences in their lengths, compositions, functions, and structures. These differences will be explored in this chapter, demonstrating how various RNAs contribute to the regulation of gene expression.

Bacterial mRNAs are translated into proteins as they are being transcribed

Bacterial mRNA, messenger RNA, is transcribed and then used immediately. Unlike eukaryotic mRNA, it is not packaged and it is not transported to another part of the cell. As a result, the opportunity for mRNA to interact with itself or with other molecules within the cell is reduced. Bacteria engage in coupled transcription–translation, which means that as the DNA is being transcribed to make mRNA, that mRNA is being translated to make the protein encoded by the gene on the DNA. There are very few changes to the mRNA transcript after it is made.

As a result, the mRNA transcript is ready to be translated into protein immediately (**Figure 4.1**). Transcription is initiated and the mRNA elongated. Before this process is complete and transcription can be terminated, translation can begin. The ribosome can initiate translation as soon as the ribosome binding site is transcribed into mRNA. DNA is transcribed to mRNA and is translated to protein together in a coupled process, which is possible because bacterial cells do not process their mRNA or transport it away from the DNA before translation.

Figure 4.1 Coupled transcription–translation in bacteria has protein being translated as mRNA is transcribed. Transcription is initiated from a promoter (bent arrow) 5′ of co-transcribed genes. Once the ribosome binding site (RBS) for the first gene is represented in the mRNA, a ribosome can initiate translation into protein. RNA polymerase (RNAP) continues elongating the mRNA through the first gene and into the second. Once the ribosome binding site for the second gene is present on the mRNA, a ribosome can bind and protein can be made from the second gene as well. As one RNAP continues its elongation, another RNAP can transcribe more mRNA. Likewise, as one ribosome moves on from the RBS to translate the mRNA into protein, another ribosome can translate the same mRNA, generating more protein.

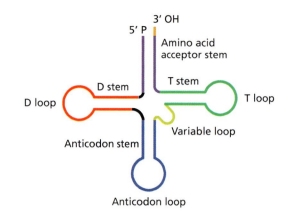

Figure 4.16 The structure of tRNAs. There are three stem loop structures in a tRNA, and optionally an additional variable loop. Starting from the 5′ end of the tRNA, these are the D-stem, D-loop, anticodon-stem, anticodon-loop, the optional variable loop, the T-stem, and the T-loop. The fourth stem, the amino acid acceptor stem, is formed by the beginning (5′) and end (3′) of the tRNA strand.

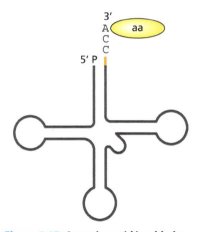

Figure 4.17 An amino acid is added to the adenine at the 3′ end of the tRNA by an aminoacyl-tRNA synthetase. The end of the tRNA contains a CCA motif, which may be part of the transcribed tRNA or it may be added later. The amino acid is added to either the 2′ or 3′ hydroxyl group of the adenine.

Figure 4.18 The tertiary structure of tRNA forms an L-shape. The two arms of the L-shape are made by the stacking of the stems, with the D-stem stacking on the anticodon stem and the amino acid acceptor stem stacking on the T-stem.

the sequence of the tRNA. Stems are formed when self-complementary regions of RNA hydrogen bond and form a region of double-stranded RNA. Loops are formed by the region between the two complementary regions. The first stem of the tRNA, starting from the 5′ end of the strand, is about four nucleotides and is referred to as the **D-stem** and the corresponding loop is the **D-loop**. Next is the **anticodon stem**, made of about six nucleotides and the **anticodon loop**, which includes the three base anticodon. The final of the three stem loop structures is the **T-stem**, of about five nucleotides, and the **T-loop**. In some tRNAs, there can be an additional **variable loop** between the anticodon stem and the T-stem (**Figure 4.16**).

The 5′ and 3′ ends of the tRNA strand pair together to form a fourth stem in the structure, the **amino acid acceptor stem**, made of about seven nucleotides. It is this fourth stem, made up of the ends of the tRNA, which accepts the amino acid, which the tRNA will carry to the ribosome during translation. There are more than 20,000 mature tRNA molecules within a bacterial cell and many of these are loaded with the corresponding amino acid. This abundance of charged tRNAs makes the molecules readily available to the ribosomes for translation. The 3′ end of the tRNA sequence contains a CCA motif, which is vital for the attachment of the amino acid. The adenine within this motif is where the amino acid attaches to the tRNA (**Figure 4.17**). **Aminoacyl-tRNA synthetases** add the amino acid to either the 2′ or 3′ hydroxyl group of the 3′ terminal adenine, which is at the end of the CCA motif. In some cases, the CCA is added to the tRNA post-transcriptionally, however, this varies by both the species and also by the tRNA.

The anticodon on the anticodon loop of the tRNA interacts with the codon on the mRNA within the ribosome. The anticodon also interacts with the aminoacyl-tRNA synthetase to ensure that the correct amino acid is loaded onto the end of the tRNA. This process ensures that during translation, the correct amino acid is added to the growing peptide chain, based on the codons encoded in the gene. The features of tRNAs vary slightly, depending on the anticodon and the amino acid that they bind. For example, the extra variable loop is only present in tRNAs for leucine, serine, and tyrosine.

The tertiary structure of tRNA is L-shaped. This shape is achieved through the stacking of the stem loop structures. The D-stem stacks on the anticodon stem. The amino acid acceptor stem stacks on the T-stem. These stacks of stems form the two arms of the L-shape. The bend in the L-shape is made of the D-loop and the T-loop (**Figure 4.18**).

The tRNA has to be properly folded to functionally interact with the ribosome, therefore, the stems of the tRNA must have bases that are able to hydrogen bond to each other. The anticodon on the anticodon loop must remain conserved; mutations here can cause misincorporation of amino acids. There are also other nucleotides within the tRNA that are conserved based on the anticodon sequence. Additionally, some nucleotides within the anticodon loop are important for tRNA interactions with the ribosome.

Figure 4.19 Some tRNAs contain the modified nucleotide pseudouridine. Pseudouridine (left) is similar to uracil (right), however, the bond between the ribose sugar and the base is a carbon−carbon (C−C) bond rather than a carbon−nitrogen (C−N) bond.

tRNA transcripts undergo post-transcriptional processing

The transcription of genes into mRNA begins before the start of the gene and extends after its end. In the same way, tRNA transcription from the DNA template begins before and ends after the sequence of the final mature tRNAs in the cell that interact with ribosomes. Also, the loci encoding tRNA are often polycistronic with other tRNAs; the different tRNAs are transcribed together and then divided into separate tRNAs by RNases. Some of the regions of the bacterial genome that encode tRNAs are near the rRNA loci.

Most tRNAs are contiguous sequences, where the sequence that is encoded in the DNA is the same as the final sequence in the mature tRNA. Some bacteria, however, have **group I introns** within tRNA sequences, as well as within genes and rRNAs. Group I introns have ribozyme activity, which means that they are capable of self-splicing to remove the region of sequence that is not present in mature tRNA.

The initial transcript, the pre-tRNA, must be processed to become a mature tRNA. Both of the ends must be trimmed, removing the flanking regions that are part of the transcript. If it is not already present in the sequence of the tRNA, the CCA motif must be added by **nucleotidyl transferase** so that the amino acid can be added to the amino acid acceptor stem by aminoacyl-tRNA synthetases. If they are present, introns are removed.

RNase P cleaves pre-tRNA. This RNase is a site-specific endoribonuclease, meaning that based on a recognition sequence within the pre-tRNA, the enzyme breaks the phosphodiester bond within the RNA cleaving the pre-tRNA. RNase P removes the 5′ leader sequence from the transcribed tRNA molecule. The RNase P enzyme contains a single RNA and a single protein that act together to recognize the RNA target and cleave it. The RNA−RNA interaction between the tRNA and the RNA within RNase P leads to cleavage.

Some of the nucleotides within the tRNA may be modified post-transcriptionally. These can include the conversion of adenine to **pseudouridine** (ψ), the conversion of adenine to **inosine** (I), and the conversion of uracil to **dihydrouridine** (D). Modifications enhance the structural stability of the tRNA, particularly in bacteria that encounter extremes or changes of temperature, and aid in correct folding. Pseudouridine is commonly found in tRNAs. It is chemically similar to uracil, however, the connection of the base to the ribose is a carbon−carbon bond, rather than the typical carbon−nitrogen bond seen in the standard nucleotides (**Figure 4.19**). Found only in tRNAs, inosine has a hypoxanthine attached to the 1′ carbon of its ribose ring (**Figure 4.20**). Dihydrouridine is formed when two additional hydrogens are added to uracil, which saturates the pyrimidine ring such that there are no longer any double bonds in the ring structure (**Figure 4.21**). Dihydrouridine can be found in tRNAs and rRNAs.

Figure 4.20 Some tRNAs contain the modified nucleotide inosine. Inosine is made when a hypoxanthine is attached to a ribose sugar.

Dihydrouridine   **Uracil**

Figure 4.21 **Some tRNAs contain the modified nucleotide dihydrouridine.** Dihydrouridine is similar to uracil, however, the addition of two hydrogens means there are no longer any double bonds in the pyrimidine ring structure.

rRNAs are essential components of the ribosome

Ribosomes are made up of both proteins and rRNAs. The RNA components are essential for the function of the ribosome. There are three types of rRNA in bacterial cells: 5S rRNA; 16S rRNA; and 23S rRNA. All three fold into complex three-dimensional shapes and then combine with ribosomal proteins to form the ribosome. Typically, the 5S rRNA, 16S rRNA, and 23S rRNA are co-transcribed as one unit. There is often more than one locus in the genome for the rRNAs. For example, *E. coli* has seven rRNA loci.

The sequences for the rRNAs are about 120 nucleotides for 5S rRNA, about 1,500 nucleotides for 16S rRNA, and about 2,900 nucleotides for 23S rRNA. These sequences are transcribed into single-stranded RNAs, which fold into complex secondary structures through the pairing of nucleotides. The secondary structures formed differ between bacterial species, in accord with species-specific sequence differences in the rRNAs.

The rRNAs participate in the activities of the ribosome. The 16S rRNA is within the 30S subunit of the ribosome and its 3′ end binds to the 5′ end of the mRNA at the Shine–Dalgarno consensus sequence. The 5S rRNA is part of the 50S ribosomal subunit and links the functional aspects of the ribosomal P-site and A-site together, although the 5S rRNA does not make contact with the tRNAs in these sites. The 23S rRNA is also part of the 50S subunit of the ribosome and is a peptidyl transferase ribozyme.

The bacterial cell also contains noncoding RNAs that can regulate other RNAs

Noncoding RNAs (ncRNAs) do not encode proteins like mRNA, do not accept amino acids and recognize codons like tRNAs, and do not complex with ribosomal proteins like rRNAs. Typical bacterial ncRNAs range in size from 50 to 500 nucleotides. Although it was known from isolation that short RNAs existed, it was believed that short RNAs were incomplete transcripts of mRNA, tRNA, or rRNA. In 1981, Tomizawa and Itoh discovered that a short RNA of about 108 nucleotides was an **antisense RNA**, the first noncoding regulatory RNA identified. This ncRNA is complementary to a region within the ColEI plasmid DNA that is cleaved to produce the replication primer. When the antisense RNA binds to this region, replication of the ColEI plasmid is blocked. There is complementary base pairing between the antisense ncRNA and the RNA that is cleaved to produce the replication primer. Other ncRNAs have since been discovered that also act by base pairing with mRNA as antisense RNAs.

Either alone or in concert with other cellular components ncRNAs can form secondary and tertiary structures. The structure of the ncRNA can impact upon its function as much as the sequence. By examining the ncRNA sequence for regions of complementarity, a structure can be predicted. The structures that can be formed due to hydrogen bonding between complementary sequences may be different at different temperatures and under different conditions, particularly in those ncRNAs

Figure 4.22 The Hfq protein assists in the binding of an ncRNA to its targeted mRNA. Base pairing between the ncRNA and mRNA forms double-stranded RNA. This is targeted by RNases within the bacterial cell, resulting in cleavage and degradation of the mRNA.

involved in the regulation of environmental responses. Pairing of ncRNAs to mRNAs is structurally driven. Either the ncRNA, its mRNA target, or both will have a stem loop region. This makes first contact with the other molecule. At the 3′ end, ncRNAs that pair with mRNAs tend to have a hairpin followed by a stretch of poly-U sequence. This structure generates an intrinsic transcriptional terminator. The ncRNA and the mRNA both contain regions where the nucleotides are complementary, which binds these two RNA molecules together. This type of ncRNA also tends to have an Hfq binding site.

The Hfq protein is a chaperone protein that stabilizes ncRNAs that carry an Hfq binding site and works in concert with the ncRNA to inactivate mRNAs. When the two RNAs come together to make a double-stranded RNA, the binding of Hfq stabilizes the structure and promotes RNase cleavage and degradation of the target mRNA (**Figure 4.22**). The Hfq binding protein is usually required for ncRNA and mRNA interactions in Gram-negative bacteria.

Although many ncRNAs regulate expression by binding to mRNA at the RBS and preventing the ribosome from initiating translation, some bind elsewhere. The inhibition of translation by other mechanisms includes the formation of secondary structures that inhibit translation elongation or that change the stability of the mRNA. There are ncRNAs that are encoded on the opposite strand to the gene they regulate. These are clear antisense RNAs, having the reverse complement sequence of the corresponding mRNA. When the antisense RNA base pairs with the mRNA, this forms double-stranded RNA that is targeted by RNases (**Figure 4.23**). It may also be that the transcription of an antisense RNA itself interferes with the ability to transcribe the gene.

Figure 4.23 Antisense RNAs are encoded on the opposite strand of the gene they regulate. The sequence of the antisense RNA is perfectly complementary to the gene sequence that has been transcribed into the mRNA because it is generated from the same DNA template, just on the opposite strand. Base pairing between the antisense RNA and the mRNA from the gene forms double-stranded RNA, which is targeted for degradation by RNases.

Key points

- The length of bacterial mRNA is similar to the length of the gene in the DNA.

- There are regions at both the 5′ and 3′ ends of mRNA that are not translated into protein.

- Secondary structure formation in mRNA can prevent translation initiation by denying the 30S ribosomal subunit access to the Shine—Dalgarno sequence.

- Translation elongation can be slowed when there are secondary structures in the mRNA that interfere with the progression of the ribosome.

- At the 3′ of mRNA, secondary structures can slow the progress of the ribosome so that the termination codon aligns properly within the ribosomal A-site.

- Riboswitches and RNA thermometers enable an mRNA to sense changes in the cell and alter the expression of the gene(s) encoded on the same mRNA molecule.

- tRNAs are folded into tight structures, with the anticodon exposed at one end and an amino acid attached at the other.

- Three rRNAs are embedded within the ribosome and act together with ribosomal proteins during translation.

- Bacterial ncRNAs are able to influence the expression of genes through interaction with mRNA.

Terms, questions, and discussions

Key terms

Define the following terms introduced in this chapter. Check your answers using the definitions in the Glossary. These terms are also available as Flashcards on-line.

Anticodon loop	Hydrolysis	RNA thermometers
Anticodon stem	Introns	Stem loop
Antisense RNA	mRNA-like domain	T-loop
Aptamer	Non-stop decay	tmRNA
D-loop	Nuclease	Transcriptome
D-stem	Nucleotidyl transferase	trans-translation
Endonucleolytic	Polyadenylation	tRNA-like domain
Exonucleolytic	Riboswitch	T-stem
Expression platform	Ribozyme	Untranslated region
Heat-shock proteins	RNase	Variable loop

Self-study questions

Answer each question using 50–100 words or a table or labeled diagram. Advice on where to find answers to these questions is available on-line.

1. A key feature of the production of proteins in bacteria is that genes are transcribed and then translated straight away. What is this process called and what features of the bacterial cell enable it?

2. What is the average size of bacterial mRNA and what dictates the size of bacteria mRNA?

3. What is the difference between the start point of transcription and the start point of translation?

4. Why might a gene be transcribed into mRNA, but not be translated into a protein?

5. What happens when there is no termination codon before the 3′ end of an mRNA and what is the name of the process that is used to recover from this?

6. Why is RNA less stable than DNA? Draw what happens.

7. Name at least three ways that secondary structures in the mRNA can interfere with the translation of the mRNA.

8. Compare the activity of a riboswitch and those that are ribozymes, considering when one might be more advantageous than the other.

9. Polyadenylation is generally thought of in the context of eukaryotic cells, but also occurs in prokaryotes. How does polyadenylation compare between eukaryotes and bacteria?

10. Draw the basic structure of a tRNA, including and labeling all of the loops and stems.

11. What are the rRNAs in the ribosome and in which subunits are each found?

12. How does the Hfq protein interact with ncRNA to lead to the degradation of mRNA?

Discussion topics

These topics are presented for discussion in study groups, as part of class discussions, or on your own. These questions go beyond what is directly covered in this part of the book. Use the research literature and other reading to explore these topics in more depth. Tips to help prepare for topic discussions are available on-line.

13. As in Figure 4.6, an mRNA transcript that is not translated by ribosomes is available to be targeted by RNases for degradation. Explore examples where this has been demonstrated to occur, either as a result of a frame-shift mutation, point mutation, or phase variation.

14. Explore the ways in which aptamers are being designed and used in either research or potential biotechnology applications. Discuss how their small size, specific structures, and specificity of binding make them powerful tools.

15. The untranslated regions 5′ and 3′ of a gene are included in the mRNA transcript and are involved in the process of translation. Discuss the ways in which the ribosome, as discussed here, and other proteins from your literature research, interact with sequences in these untranslated regions. How do such interactions influence the expression of proteins?

Further reading

mRNAs

Del Campo C, Bartholomäus A, Fedyunin I, Ignatova Z. Secondary structure across the bacterial transcriptome reveals versatile roles in mRNA regulation and function. *PLoS Genet.* 2015; 11(10): e1005613.

tRNAs

Lorenz C, Lünse CE, Mörl M. tRNA modifications: Impact on structure and thermal adaptation. *Biomolecules.* 2017; 7(2): 35.

Shepherd J, Ibba M. Bacterial transfer RNAs. *FEMS Microbiol Rev.* 2015; 39: 280–300.

rRNAs

Petrov AS, Bernier CR, Gulen B, Waterbury CC, Hershkovits E, Hsiao C *et al.* Secondary structures of rRNAs from all three domains of life. *PLOS ONE.* 2014; 9(2): e88222.

Small RNAs

Michaux C, Verneuil N, Hartke A, Giard JC. Physiological roles of small RNA molecules. *Microbiology.* 2014; 160: 1007–19.

Tomizawa J, Itoh T. Plasmid ColE1 incompatibility determined by interaction of RNA I with primer transcript. *Proc Natl Acad Sci U S A.* 1981 Oct; 78(10):6096–100.

Transcriptional Regulation

All RNA is generated by transcription. The initiation of the process requires RNA polymerase to bind to the promoter region present in the DNA. Promoter regions are located 5′ of the region that will be transcribed into RNA. Transcriptional regulation modifies the frequency of transcription, regulating the amount of RNA made. For some RNAs, transcription occurs at a relatively consistent level. The coding regions encoded by these RNAs are said to have **constitutive expression**, which means that the genes are transcribed at roughly the same levels at all times. This does not mean that all constitutively expressed genes are expressed at the same levels relative to each other; the individual levels of expression vary depending on the gene. Constitutive expression refers only to the level of transcription of the gene being equal regardless of the state of the cell.

Genes encode different proteins, which may be required at different levels within the cell at different times. The amount of mRNA transcribed from some genes is therefore different at different times, depending on the state of the cell. Many genes are subject to **regulated expression**, where the levels of mRNA transcribed may be higher or lower at various times for the cell and under various conditions. Regulation of mRNA levels is often at the point of transcription initiation. If RNA polymerase cannot initiate transcription for a gene, then there will be little to no mRNA made and without the mRNA carrying the message of the gene, it cannot be translated into protein. If binding of RNA polymerase to the promoter region is facilitated, then the initiation of transcription can occur more readily, producing an increase in the amount of the mRNA carrying the message encoded in the gene. Modulation in the amount of RNA synthesized is **transcriptional regulation**.

Regulation of gene expression at the level of transcription

Gene expression relies on a series of factors that must all be in place. The first is that the DNA encodes a functional copy of the gene. This gene sequence must then be transcribed into mRNA, so that it can be expressed as a protein through translation. The process of transcription can be controlled, resulting in genes being expressed when they are needed and being switched off when they are not needed. The result is that proteins are made when they are needed, rather than all of the time. By regulating mRNA production and protein synthesis, the energy of the cell is conserved.

Regulation of gene expression can be influenced by the condition of the bacterial cell. For example, the types and levels of nutrients available in the environment can change. Bacteria must respond to these changes in order to survive, making best use of the nutrients that are available. Response to a changing environment or conditions can result in alterations to the genes expressed at the level of transcription.

Transcriptional regulation of genes is often complex, requiring multiple components. The sigma subunit of RNA polymerase must be able to recognize the promoter region to initiate transcription. Different sigma factors will recognize different promoter region sequences. In addition, there may be multiple DNA-binding proteins with separate, overlapping, or the same binding sites within or proximal to the promoter. The promoter region can also have binding sites for noncoding RNAs. Genes can be part of regulatory cascades, where activation or repression of transcription of one gene impacts upon the expression of another

Figure 5.1 The breakdown of lactose. Within a bacterial cell such as *E. coli*, the enzyme β-galactosidase can break down the sugar lactose into its component sugars galactose and glucose.

gene and so on down the chain. Proteins or ncRNAs may bind to the transcript itself, altering the progression of its transcription or the translation of the mRNA. The structure of the genome and the associated availability of promoter regions can impact the ability for RNA polymerase to access the DNA, influencing the initiation and completion of transcription.

The classic example of transcriptional regulation: The *lac* operon

The first example of transcriptional regulation was described by Jacques Lucien Monod and François Jacob and remains the classical example of the mechanisms involved in changes in transcription as a result of cellular conditions. In 1961, Monod and Jacob published a study describing their findings. They made a connection between the binding of proteins to DNA and the regulation of the amount of protein within the cell. It was only later that mRNA was shown to be the link between the DNA and its encoded protein.

The regulation of gene expression in the classic *lac* operon system was investigated in *Escherichia coli*. Metabolically, *E. coli* and other bacteria require sugars to generate cellular energy stored as ATP. The disaccharide **lactose** is broken down within the bacterial cell by the enzyme β-**galactosidase** to its component sugars **glucose** and **galactose** (**Figure 5.1**). The β-galactosidase enzyme also converts lactose into the metabolite allolactose. Lactose is transported into the bacterial cell by the transport protein β-**galactoside permease**. The genes encoding β-galactosidase and β-galactoside permease are adjacent to one another in the DNA of *E. coli*, encoded by *lacZ* and *lacY*, respectively. These genes are followed by *lacA* in a single transcriptional unit. Upstream of the three *lac* operon genes, collectively referred to as *lacZYA*, is the *lacI* gene, which encodes a **repressor protein**. Between the *lacZYA* and *lacI* genes is the promoter for the *lacZYA* genes and an **operator**, which is a binding site for proteins to bind to DNA and influence transcription (**Figure 5.2**). The *lacZYA* genes make up the *lac* **operon** and the associated features are involved in regulation of the *lac* operon. Operons are co-transcribed genes that are also co-regulated, such that these adjacent genes are transcribed together by RNA polymerase and are together regulated.

When there is no lactose in the *E. coli* cell, the repressor protein LacI binds to the operator, preventing the transcription of *lacZYA*. Without transcription of the genes, the proteins that they encode are not made. When the *E. coli* cell does contain lactose, the *lacZ* encoded β-galactosidase converts some of the lactose to allolactose, which binds to the LacI repressor protein. There is a conformational change in the LacI repressor protein when it binds allolactose, which causes the protein to no longer be able to bind to DNA. In this capacity, lactose as allolactose is an **effector molecule**, a cellular component that causes or prevents transcription. If mutations are made in the DNA such that there is no LacI repressor protein, then the *lacZYA* genes are always on, becoming constitutively expressed. Likewise, alteration or deletion of the operator sequence results in constitutive expression. If there is no promoter, then RNA polymerase has no binding sequence with which to initiate transcription and no mRNA carrying the sequences of the *lacZYA* genes is made. Because the *lacZ* gene encodes the β-galactosidase,

Figure 5.2 The structure of the *lac* operon. The *lacZ*, *lacY*, and *lacA* genes form one transcriptional unit. Upstream is the *lacI* gene encoding LacI, the repressor of transcription of *lacZYA*. The LacI repressor protein binds to a region between the *lacI* and *lacZYA* genes called the operator (Op). Binding of LacI to the operator prevents RNA polymerase from binding to the promoter (bent arrow) and initiating transcription of *lacZYA*.

deletion of this gene means that lactose cannot be broken down into glucose and galactose and no allolactose is produced. A deletion of the *lacY* gene encoding the β-galactoside permease prevents the entry of lactose into the bacterial cell.

The *lac* operon is also subject to catabolite repression

The *E. coli* cell will use glucose by preference over other sugars. Capture of useable energy, mainly as ATP, is more efficient from glucose than from lactose or other sugars and thus the cell will use glucose first over other sugars available. The preference for certain molecules causes **catabolite repression**, where in this case the presence of glucose represses the expression of genes needed for the metabolism of lactose.

In addition to the operator binding site for LacI, the promoter region for *lacZYA* contains a binding site for CAP–cAMP. **CAP** is the **catabolite activator protein**, which is able to form a complex with **cyclic AMP (cAMP),** a form of adenosine monophosphate where the phosphate is bound to both the 5′ and 3′ carbons (**Figure 5.3**). The binding of the combined CAP–cAMP to its *lac* operon binding site on the DNA is required for the transcription of *lacZYA*. While the LacI repressor has a negative influence on transcription, the binding of CAP–cAMP to the promoter region exerts positive control upon the *lac* operon (**Figure 5.4**). The binding of CAP–cAMP to DNA is associated with the presence of glucose within the bacterial cell. When the concentration of glucose is up, the concentration of cAMP is down. This is because glucose is efficiently processed to produce ATP. Because cAMP is scarce, it does not complex with the protein CAP and there is no binding of CAP–cAMP to activate transcription of the *lac* operon.

If glucose is at low levels in the cell, but there is no lactose available, then the *lac* operon is off. It is not transcribed because the correct combination of control elements that would permit transcription of *lacZYA* is not present. When glucose is low, cAMP is high and is able to bind to CAP, which together are able to bind to the operator in the promoter region of the *lac* operon. This should allow the initiation of transcription, however, without lactose in the cell the LacI repressor protein is able to bind to its operator, preventing transcription. For there to be transcription of the *lacZYA* genes there must be lactose, as allolactose, present to bind to the LacI repressor protein and change its conformation, otherwise, the repressor will bind to the operator and negate the positive influence of the binding of CAP–cAMP. When glucose is low or absent and lactose is present, the allolactose binds to the LacI repressor protein and the conformational change caused eliminates the DNA binding function of the repressor. Since low glucose levels result in abundant cAMP, there is plenty of it to bind to CAP. The CAP–cAMP complex is then able to bind to the operator in the *lac* operon and encourage the initiation of transcription of *lacZYA* by RNA polymerase. Therefore, when there is no glucose, but there is lactose, the enzymes that will enable the *E. coli* to use lactose are made. The two levels of control on the *lac* operon, which senses glucose and lactose availability, stops the bacterial cell wasting energy on making unnecessary enzymes for lactose metabolism when glucose is available and when there is no lactose (**Figure 5.5**).

Cyclic-AMP

Figure 5.3 The structure of cAMP. AMP is adenosine monophosphate, which is generated when ATP, adenosine triphosphate, loses two phosphates and releases the energy for those bonds. Cyclic AMP (cAMP) is a form of AMP where the remaining phosphate is bonded to both the 5′ and the 3′ carbons on the ribose sugar.

Figure 5.4 The binding of CAP–cAMP to the *lac* operon. When glucose is scarce, cAMP is abundant and binds to the catabolite activator protein (CAP). This CAP–cAMP complex is able to bind to the intergenic region between *lacZYA* and *lacI*. This binding is required for the expression of *lacZYA* from its promoter (bent arrow).

Figure 5.5 The regulation of the *lac* operon. When glucose is low and lactose is available (a), allolactose binds to the LacI repressor, preventing it from being able to bind to the operator. At the same time, cAMP is available to bind to CAP, resulting in the CAP—cAMP binding to the DNA and turning on transcription of the *lac* operon by RNA polymerase (RNAP). When glucose is high (b and d), cAMP is low and the CAP—cAMP complex is not made, so the positive signal for *lac* operon transcription is absent. If there is no lactose present (b and c), the LacI repressor protein will bind to the operator and prevent any transcription if the *lac* operon, even when glucose is low and CAP—cAMP binds to the DNA (c). In a cell high in glucose where lactose is also present (d), the *lac* operon is transcribed, but at a low level in the absence of CAP—cAMP binding.

The actions of the corepressor tryptophan on the *trp* operon

Tryptophan is an amino acid required for the translation of some proteins. Tryptophan can either be acquired from the environment or if there is no tryptophan available, it can be synthesized by the bacterial cell. When the concentration of tryptophan within the bacterial cell is low, there is transcription of the *trp* operon. When the concentration of tryptophan is high, transcription of the *trp* operon is repressed. There are five co-transcribed genes in the *trp* operon, which encode enzymes that are able to produce tryptophan for the cell when it is scarce. The five genes, *trpEDCBA*, are preceded by a promoter and an operator (**Figure 5.6**). It is costly to the bacterial cell to make these enzymes and to produce tryptophan, so the *trp* operon is turned off when the cell can obtain sufficient tryptophan from its environment.

Regulation of the *trp* operon is via a repressor protein that is able to bind to tryptophan and block the transcription of the genes. In this system, tryptophan acts as a **corepressor**, which is a small molecule that is able to switch a repressor protein to its active DNA-binding conformation by binding to the protein. Transcription of

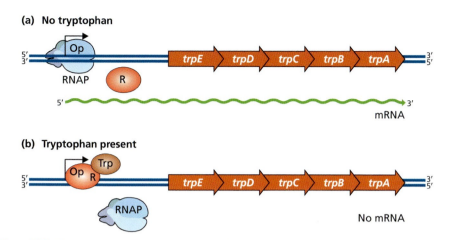

(a) No tryptophan

(b) Tryptophan present

Figure 5.6 The structure of the *trp* operon and its transcription. The five genes of the *trp* operon, *trpEDCBA*, encode enzymes that are able to make tryptophan for the cell. These genes are transcribed from the promoter (bent arrow), when there is no tryptophan available to the bacterial cell (top). However, when tryptophan is present a repressor protein (R) is able to bind to the operator (Op) in concert with the corepressor tryptophan (Trp) and prevent transcription by RNA polymerase (RNAP).

the *trp* operon is not initiated in the presence of tryptophan, due to the action of the repressor protein and the corepressor, tryptophan.

An attenuation mechanism controls the expression of the *trp* operon

The *trp* operon is also subject to regulation by a system called **attenuation**. This process prevents the formation of a complete strand of mRNA from the *trp* operon. When tryptophan is abundant, RNA polymerase stops transcription prematurely. Between the operator and the *trpE* gene, there is a region called the **leader** followed by another region called the **attenuator**. The leader sequence encodes a short polypeptide, which is made from the leader sequence mRNA transcript. The attenuator sequence contains regions of self-complementary nucleotides and as such forms hairpins in the mRNA transcript. The types of hairpins that are formed by the attenuator depend on the translation of the leader. The influence of mRNA secondary structure on gene expression was discussed in more general terms in Chapter 4 (see Figure 4.7).

RNA polymerase, in the absence of the *trp* operon repressor protein, binds and initiates transcription of the *trpEDCBA* genes, as well as the leader and attenuator sequences that precede the genes (**Figure 5.7**). Shortly after the initiation of transcription, a ribosome binds to the mRNA and initiates translation. The first region to be translated is the leader sequence, which encodes the 14 amino acid leader peptide. The leader sequence includes two codons for tryptophan amino acids, therefore, during translation of the leader, the ribosome must incorporate two tryptophans into the leader peptide. If there is a high concentration of tryptophan in the bacterial cell, the ribosome is able to translate the leader peptide quickly, rapidly

Figure 5.7 The features of the *trp* operon mRNA transcript. The mRNA transcribed from the *trp* operon includes, from 5′ to 3′, a leader sequence, an attenuator sequence, *trpE*, *trpD*, *trpC*, *trpB*, and *trpA*. The leader sequence is a short coding region, which encodes a peptide of 14 amino acids, including two tryptophans. The attenuator sequence contains regions of self-complementarity, where the mRNA forms hairpin secondary structures (see Figure 5.8). The *trpEDCBA* genes encode proteins capable of synthesizing tryptophan.

Figure 5.8 The secondary structures of the 5′ end of the *trp* operon mRNA. (a) The mRNA of the *trp* operon includes a leader sequence at the 5′ end that encodes a short peptide, followed by an attenuator sequence. There are four complementary regions that can base pair with one another forming secondary structures. (b) In one pairing, the leader sequence-containing region 1 pairs with region 2 and region 3 pairs with region 4. The pairing of 3 and 4 forms a transcriptional terminator. (c) Alternatively, regions 2 and 3 can pair together, forming an anti-terminator.

forming the polypeptide and dissociating from the mRNA. If the levels of tryptophan within the cell are low, the ribosome stalls while it waits for tRNAs carrying tryptophan to arrive at the A-site and continue translation. The leader polypeptide is therefore made slowly when there is little tryptophan present.

The attenuator sequence, at the 5′ end of the *trp* operon mRNA transcript, forms secondary structures in the mRNA. The attenuator region can form two different configurations of hairpins (**Figure 5.8**). One conformation forms a transcriptional termination signal. The other conformation does not form a transcriptional termination signal, rather it stops the transcriptional terminator hairpin from forming. This mRNA hairpin is an **anti-terminator**. If the ribosome is slow to progress through the leader sequence codons due to the scarcity of the tryptophan amino acid, it pauses and the pausing allows the formation of the anti-terminator hairpin and results in transcription of the complete *trp* operon mRNA (**Figure 5.9**). If tryptophan is abundant within the bacterial cell, the ribosome is able to quickly progress with translation. Upon the completion of translation of the leader sequence, the ribosome dissociates from the mRNA. This rapid progression of the ribosome allows the formation of the transcriptional termination hairpin and of a second hairpin in the attenuator sequence. These hairpins in the mRNA cause RNA polymerase to come off from the DNA, terminating transcription prematurely

Figure 5.9 The translation of the *trp* operon leader polypeptide and anti-terminator formation. Within the 5′ end of the *trp* operon mRNA, there is a short leader sequence encoding a polypeptide containing two tryptophan amino acid codons. When there is little tryptophan in the cell, the ribosome translating the leader must pause to wait for tRNA^{Trp}. This pausing allows the adjacent attenuation sequence to form an anti-terminator hairpin. RNA polymerase (RNAP) continues to transcribe the *trp* operon.

Figure 5.10 The translation of the *trp* operon leader peptide and transcriptional terminator formation. Within the 5′ end of the *trp* operon mRNA, there is a short leader sequence encoding a polypeptide containing two tryptophan amino acid codons. When tryptophan is abundant within the bacterial cell, the ribosome translating the leader will progress quickly. This rapid translation enables the attenuation sequence to form a transcriptional terminator hairpin. As a result, RNA polymerase (RNAP) ceases transcription of the *trp* operon and dissociates from the DNA.

(**Figure 5.10**). In this configuration of hairpins, a complete transcript for the *trp* operon is not made, negatively regulating the expression of the tryptophan synthesis genes when tryptophan is abundant.

When tryptophan in the cell is high, the ribosome is quick through the leader, the transcriptional terminator forms, and transcription of the *trp* operon ends prematurely. When tryptophan is low, the ribosome is slow, the anti-terminator forms, and there is transcription of the complete *trp* operon, enabling tryptophan to be synthesized by the cell.

Genes are regulated locally by *trans*-acting factors

The organization of genes into operons such as the *lac* operon and the *trp* operon facilitates the synchronous expression of these genes. However, distant genes can also be co-regulated through the use of the same regulatory proteins to orchestrate their expression. While the regulator protein for the *lac* operon is encoded near to its binding site for repression of *lacZYA*, regulatory proteins can be encoded anywhere in the genome and can bind anywhere in the genome. Although many regulatory proteins may act on transcription of genes near to where they are encoded on the chromosome, they can also act in *trans*, influencing the expression of distant genes. For example, in addition to the genes of the *lac* operon, transcriptome studies have shown that *E. coli* that are grown in the presence of lactose and absence of glucose have altered the regulation of hundreds of other genes in the genome.

As seen in the *lac* and *trp* operons, there are DNA binding proteins that are repressors of transcription and those that are activators of transcription. Some regulatory proteins bind to a metabolite or other small molecule within the bacterial cell and this binding, or lack thereof, changes the capacity of the regulatory protein to bind to DNA. Other regulatory proteins simply bind to their sequence-specific binding site on the DNA. In order to interrupt the influence of a regulatory protein that acts on its own to bind to DNA, the expression of the regulatory protein has to be stopped. This can be accomplished either by preventing the gene encoding the regulator from being transcribed or prevent the transcribed mRNA from being translated into the protein. Influencing the transcription or translation of a regulatory protein creates a regulatory cascade, where the regulation of the level of expression of the regulatory protein impacts upon the level of expression of the genes that the regulatory protein regulates.

Repressors, activators, and inducers can influence the expression of many genes

Repressor proteins, which stop or reduce the expression of genes through binding to the DNA in or near the promoter region, can act on more than one promoter.

Several genes can be co-regulated by a repressor protein. If, for example, the cell does not need to manufacture proteins for a particular biosynthetic process, the use of a single repressor can stop transcription of all of the genes involved in the process. Regardless of their location in the genome, the repressor protein can coordinate and turn off the expression of genes when their products are not needed, reducing the waste of energy involved in making proteins that are not needed.

Often working in opposition to repressors are activator proteins. These DNA binding proteins assist in the binding and recruitment of RNA polymerase to the promoter. The actions of the activator protein result in an increased expression of the gene. In some cases, this may be through interaction between the activator protein and the RNA polymerase, where the activator bound to its sequence-specific binding site may aid in the binding of the RNA polymerase to a weak promoter region. Like repressors, activators can regulate genes that are scattered throughout the genome. Synchronized activation of genes can be achieved when a set of genes are turned on by the same activator protein. This can ensure that all genes required for a biosynthetic process are turned on simultaneously.

Both repressors and activators can be subject to the effects of inducers. Inducers are molecules within bacterial cells that can bind to regulatory proteins and influence their activity. When an inducer binds to a repressor, the inducer prevents the repressor from binding to DNA and repressing expression. Allolactose is an example of an inducer that interacts with a repressor. In the *lac* operon, the allolactose binds to the LacI repressor, preventing it from repressing the transcription of the *lacZYA* genes. When an inducer binds to an activator it enhances the activation ability of the activator. It does this by aiding the activator in strongly binding to the DNA. Without the inducer, these activator proteins have very poor binding to the DNA. In the *lac* operon, cAMP binds to the regulatory protein CAP, enabling it to bind to DNA and activate transcription.

Two-component regulators sense change and alter transcription

Two-component regulatory systems are one of the main mechanisms used by bacterial cells to respond to stimuli. In response to stimulus, gene expression is changed. Cellular functions involving differential gene expression such as pathogenesis, biofilm formation, and motility, can be regulated by two-component systems. There are two proteins involved in two-component regulatory systems; these two components give the system its name. The first protein component is a **sensor kinase**, which is a membrane-embedded receptor protein that recognizes the signal for the two-component regulatory system. When this sensor kinase binds to its ligand, this causes the sensor kinase to **autophosphorylate**, adding a phosphate from ATP to a specific histidine within the sensor kinase itself. The phosphorylated form of the sensor protein is able to transfer its phosphoryl group to the second protein of the two-component regulatory system. This second protein is the **response regulator** protein, which is only active when it is phosphorylated. The phosphate from the sensor kinase is transferred to an aspartic acid on the response regulator protein. This changes the conformation of the protein, so that the active response regulator is able to bind to DNA at the promoter of the genes it regulates and either increase or decrease transcription (**Figure 5.11**).

Phosphorelays are similar to two-component regulatory systems, however, the sensor kinase does not directly phosphorylate the protein that will bind the DNA and

Figure 5.11 Two-component regulatory systems regulate transcription in response to external signals. The membrane-bound sensor kinase (SK) senses a signal from outside the cell and autophosphorylates, using a phosphate from ATP. This phosphoryl group is then passed on to a response regulator (RR). The phosphorylated response regulator in the two-component system undergoes a conformational change, enabling it to bind DNA and regulate gene expression in response to the external signal.

regulate expression. Instead, there is an expanded cascade involving four proteins, along which the phosphate is passed. After autophosphorylation in response to stimuli, the sensor kinase transfers its phosphoryl group to a response regulator that has no DNA binding ability, even when activated. This response regulator transfers the phosphoryl group to a phosphotransfer protein, which in turn transfers it to the terminal response regulator that binds DNA (**Figure 5.12**). In some cases, the sensor kinase and the first response regulator are fused into one protein, embedded in the membrane.

DNA changes in the promoter region locally regulate transcription in *cis*

Transcription relies upon DNA, not only to be the template for the transcription of mRNA, but also to contain all of the sequence signals needed to initiate transcription. The sequence information is able to therefore exert a *cis* effect upon the mRNA.

The clearest example of this is the promoter itself. Without a promoter, there is no sequence for the sigma factor to recognize and recruit RNA polymerase to initiate transcription. Therefore, any mutations in the promoter sequence DNA may impact the level of transcription by altering recognition of the promoter by the sigma factor (**Figure 5.13**). There may also be more than one promoter upstream of a gene and the location of these sequences in the DNA may influence which promoter is used to drive expression, when transcription is initiated in the cell, and at what level the region is transcribed.

Operators also exist within the DNA. These sequences are bound by regulatory proteins and changes to their sequences may alter the binding of the DNA binding proteins (**Figure 5.14**). The binding of regulatory proteins to their binding sites is sequence specific. The protein has to be able to recognize its binding site to achieve effective binding and either prevent transcription by RNA polymerase or enhance it. Likewise, any changes to RNA binding sites, where regulatory RNAs will bind to the mRNA transcript, can change the expression level of the associated gene.

Programmed changes to DNA can alter transcription locally

Some mutations within promoter regions are programmed into the DNA itself. Due to the sequence features present, the level of expression from the promoter changes based on the promoter mutations. One mechanism for this is through changes in

Figure 5.12 Phosphorelays regulate transcription in response to external signals. The membrane-bound sensor kinase (SK) in the phosphorelay system senses a signal from outside the cell and autophosphorylates, using a phosphate from ATP. This phosphoryl group is then passed on to a response regulator (RR) that lacks a DNA binding domain. The phosphoryl group is passed on again, this time to a phosphotransfer protein (PTP), which transfers the phosphoryl group to a second response regulator (RR). This last protein in the phosphorelay gains the ability to bind DNA following a phosphorylation-induced conformational change.

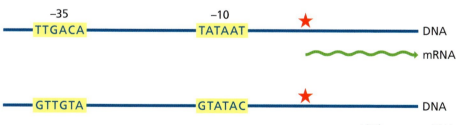

Figure 5.13 Promoter mutations alter the expression of genes. Mutations in the consensus promoter region recognized by the sigma factor can impact the binding of the sigma factor and thus RNA polymerase. When the −10 and −35 consensus promoter sequences are present, σ^{70} can bind to the promoter, recruit RNA polymerase, and initiate transcription of mRNA from the start point of transcription (★) (top). If a mutation reduces or eliminates binding of the sigma factor to the promoter, little to no transcript of the gene is made by RNA polymerase and the gene is not expressed. When the sequence is too different from the recognized consensus sequence (bottom), there will be no mRNA produced because transcription cannot be initiated.

Figure 5.14 Binding of regulatory proteins to DNA is sequence specific. The regulatory protein will bind to its binding site based on the DNA sequence (top). If there are changes in the DNA sequence, it may no longer be recognized by the regulatory protein, which will not bind (bottom). This can impact the expression of the associated gene either through lack of repression or lack of activation.

Figure 5.15 Changes in simple sequence repeats alters the promoter and causes phase variation. In bacterial species that use transcriptional phase variation, simple sequence repeats are associated with promoter regions. Changes in the length of these repeats alters the promoter, which may result in increased or decreased ability for the sigma factor to bind to its consensus, recruit RNA polymerase, and initiate transcription at the start point (★). Here the number of G bases between the −10 and −35 promoter sequences has changed. The number of G bases between the −10 and −35 promoter regions of the DNA creates ideal spacing of the promoter elements and RNA polymerase is able initiate transcription and produce mRNA (top). A reduction in the number of G bases creates a promoter region where the −10 and −35 sequences are not ideally spaced, so RNA polymerase cannot recognize the promoter and no mRNA is made.

simple sequence repeats. These can be homopolymeric tracts of the same base or they can be repeated short sequences of bases. Changes in the lengths of these repeats are introduced during replication and impact upon the expression of the gene from the repeat associated promoter (**Figure 5.15**). This process is called **phase variation** and is used by some bacteria to generate diversity in gene expression within the population of bacteria.

Other mechanisms may alter the DNA upstream of a gene and influence the expression of that gene. There are some genes in some genomes that are driven by promoters that are affected by inversion events, such as *fimA* in *E. coli* involved in fimbriae expression. These are not random mutations, but mechanisms programmed into the DNA itself. Localized changes to the DNA, where a segment of DNA within the genome is excised, inverted, and reinserted at the same location, can change the orientation of a promoter so that it is no longer upstream of the gene. The inversion event may switch between a gene having a promoter and being expressed or not having a promoter and not being expressed (**Figure 5.16**). Promoter inversion may also alter the strength of the expression level, switching between a strong promoter in one orientation and a weak one in the other orientation, or switching between two promoters recognized by different sigma factors.

The DNA of bacterial cells can be methylated and these methyl groups may disrupt the ability of regulatory proteins to bind to their binding sites. **Methylation** in bacteria occurs at the C-5 or N-4 of cytosine or at the N-6 of adenine when a methyl

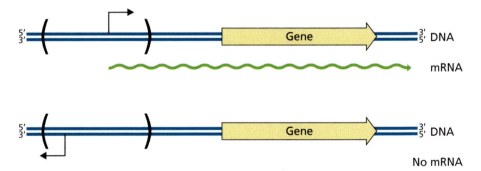

Figure 5.16 Gene expression may change due to inversion events. Inversion of a region of DNA, where the sequence is turned around in the opposite orientation, can impact upon expression. Promoters in the invertible region (marked with parentheses) are switched in their orientation relative to the gene. The inversion has switched the expression of the gene, with mRNA being produced when the promoter is in one orientation (top) and not when it is in the other (bottom).

Figure 5.17 Methylation of DNA can influence the binding of regulatory proteins. The methylation state of DNA can change the binding of regulatory proteins, such as the binding of Lrp and PapI to the promoter region of the pilus genes *papAB* in *E. coli*. Some bases of DNA can become methylated, here adding a methyl group to an adenine. The presence or absence of the methylation can alter the capacity of regulatory proteins to recognize the binding site, bind, and modify the expression of the gene by RNA polymerase. When GATC^{prox} is methylated, Lrp and PapI bind to the GATC^{dist} region, allowing RNA polymerase to recognize the promoter (bent arrow) and initiate transcription (top). When GATC^{dist} is methylated, the Lrp regulatory protein binds instead to the GATC^{prox} region repressing transcription by preventing RNA polymerase from accessing the promoter (bottom).

group is added by a methyltransferase from *S*-adenosyl methionine. The methylation of an adenine or cytosine can alter the binding of regulatory proteins to the DNA at that site, either due to the physical presence of the methyl group or due to a change in the structure of the DNA that influences binding. The binding of proteins to DNA can also block the methylation of the DNA by methyltransferases.

In *E. coli*, expression of pili can switch on and off as a result of changes in methylation at the promoter for the pilus protein. There are two GATC methylation sites, GATC^{prox} and GATC^{dist}, where the adenine can be methylated by DNA methylases. These two methylation sites overlap the binding sites for the Lrp regulatory protein and the PapI regulatory protein. When GATC^{prox} is methylated but GATC^{dist} is not, the gene expression of *papAB* encoding the pili is on. PapI and Lrp bind to the DNA at GATC^{dist} and activate transcription. However, when GATC^{dist} is methylated and GATC^{prox} is not, expression of pili is off. Lrp binds to GATC^{prox} and represses transcription by preventing RNA polymerase from accessing the promoter (**Figure 5.17**).

Sigma factors are essential for the initiation of gene transcription

Sigma factors function by binding to promoters and combining with the RNA polymerase core enzyme to make the RNA polymerase holoenzyme. The RNA polymerase holoenzyme, made of the core enzyme and a sigma factor, is capable of initiating transcription. The availability and function of sigma factors may be altered by environmental conditions. The sigma factors, in addition to their role in initiating transcription, can therefore regulate transcription.

Genes can have two or more promoters, recognized by different sigma factors. There is a great deal of diversity within bacterial species, both in terms of the numbers of different sigma factors that the cell can make and also the specificity of each of the sigma factors. Sigma factors each recognize a distinct, specific sequence within the promoter and bind to them. Different sigma factors therefore regulate the expression of genes through recognition of different promoters.

After the initiation of transcription, the sigma factor leaves the RNA polymerase holoenzyme and the RNA polymerase core enzyme continues with transcriptional elongation. This sigma factor is then able to go on to a promoter and initiate transcription by joining with another RNA polymerase core enzyme. There is a limit to the amount of RNA polymerase in the bacterial cell, therefore, the sigma factors compete with one another for a limited pool of RNA polymerase core enzyme with which to initiate transcription.

Sigma factors can orchestrate global regulation of gene transcription

Many genes are transcribed by RNA polymerase with the σ^{70} housekeeping gene sigma factor. Alternative sigma factors are able to initiate transcription of a variety of genes, often enabling the bacteria to survive stresses, respond to nutrient availability, and orchestrate morphological changes. Alternative sigma factors are often involved in survival in stress, which can be survival in hosts. Within the host, the expression of bacterial virulence factor genes and within host survival genes can be controlled by alternative sigma factors, such as σ^S involved in stress responses and σ^{32} involved in temperature responses. For example, the ECF sigma factor, for extracytoplasmic functions, is involved in the initiation of transcription of genes in response to cell envelope stress and the environment.

The promoters of heat-shock proteins are recognized by σ^{32}, the key regulator of the heat-shock response. The half-life of σ^{32} at the normal temperature for a bacterial species is about 1 minute, but at higher temperatures the half-life increases to about 5 minutes. Therefore, at higher temperatures σ^{32} will stay in the cell longer and be able to participate in the **initiation** of transcription more often than at normal temperature. Once sufficient quantities of heat-shock proteins have been made, the activity and quantity of σ^{32} in the cell can be reduced. Heat-shock proteins DnaK, DnaJ, GrpE, and HflB are chaperones to misfolded proteins, however, as the number of misfolded proteins decreases, these heat-shock proteins instead interact with σ^{32} and target it for degradation. The half-life of the sigma factor and its interaction with heat-shock proteins that it up-regulates work together to optimize the amount of σ^{32} in the bacterial cell based on temperature and protein misfolding.

Control of sigma factor activity involves several components

Control of sigma factor expression can involve small noncoding RNAs. In some cases, these act at the level of translation, including translation from existing mRNA as well as newly synthesized mRNA. For example, generation of sigma factor protein σ^S (RpoS) is driven by stresses. The *E. coli* σ^S has about 500 genes for which it controls expression. The σ^S itself is regulated by at least four ncRNAs and requires the RNA chaperone Hfq. One ncRNA, DsrA, senses changes in temperature, another, RprA, senses stress at the cell surface, and a third, OxyS, responds to oxidative stress. Each increases translation of the σ^S encoding mRNA. The ncRNAs DsrA and RprA, together with Hfq, bind at the 5′ end of the *rpoS* mRNA, stopping the formation of a hairpin and making the RBS accessible by ribosomes. The presence of σ^S within the cell is also controlled through regulation of its own transcription and differential degradation of the σ^S protein by **proteolysis**.

There are also **anti-sigma factors** that are able to prevent the action of sigma factors. Anti-sigma factors bind to sigma factors and block the portion of the sigma factor that binds to the RNA polymerase core enzyme. Often anti-sigma factors inhibit the activity of the alternative sigma factors, keeping them sequestered until they are needed. When the alternative sigma factors are released from anti-sigma factor control, anti-sigma factors may then inhibit the housekeeping sigma factor σ^{70}

to prevent it from initiating transcription. The genes for anti-sigma factors tend to be co-transcribed with those for the sigma factor that they regulate.

Global regulation can be influenced by the binding of chromatin proteins to the DNA

Bacterial DNA exists in a compact mass, known as the nucleoid. One of the ways the compaction of DNA is achieved is through DNA supercoiling, using the turns of the double helix to compact the DNA into a small size. There are **chromatin proteins** associated with the DNA within the nucleoid. These proteins also contribute to the compactness of the DNA within the cell. Macromolecular crowding of all of the bacterial cellular components within the cell also plays a role in compacting the DNA. Bacterial DNA takes up only about a quarter of the interior of the bacterial cell, which is also full of enzymes, RNAs, and other small molecules that crowd the DNA into a small space.

The transcription in some regions, such as the loci for the rRNAs and tRNAs, is high. The DNA for these cellular features must be accessible to the transcription machinery locally, yet the DNA also has to be compact overall. Therefore, the structure of the DNA, its supercoiling, interaction with chromatin proteins, and macromolecular crowding are involved in the global regulation of the features on the chromosome. The main chromatin proteins that contribute to nucleoid compaction in most bacterial cells are **H-NS** the **histone-like nucleoid structuring protein**, **HU** the **heat unstable protein**, **Fis** the **factor for inversion stimulation**, and **IHF** the **integration host factor**.

The H-NS protein binds to DNA, making regions unavailable for transcription

The structure of the H-NS protein has an N-terminal dimerization domain that enables the protein to form **dimers** with another H-NS. The C-terminal domain is the DNA-binding domain of the protein, which is connected to the N-terminus by a flexible region within the H-NS protein. H-NS proteins preferentially bind to A+T-rich regions of the DNA. When it binds to these regions it silences them. Within the bacterial cell genome, if regions of horizontal transfer are A+T-rich, having originated from a species that has lower G+C, the H-NS binding can prevent expression of foreign genes that may be detrimental to the cell.

H-NS is present in approximately 20,000 copies of the protein per cell, making it abundant within the cell and a major protein component of the nucleoid. Regulation in gene expression due to H-NS is a result of the capacity of many copies of H-NS to together bind to contiguous regions of DNA and cause it to bend through interactions between H-NS proteins (**Figure 5.18**). The H-NS induced bending prevents RNA polymerase from accessing promoter regions to initiate transcription from the DNA. In addition, the binding of H-NS to regions of DNA can block the progression of RNA polymerase during transcriptional elongation.

Figure 5.18 H-NS proteins bind to DNA, compacting it and bending it. Multiple copies of the H-NS protein are able to bind to the DNA and to each other, causing the DNA to form compact structures. The bends in the DNA created by H-NS binding brings together the nucleoid and condenses the DNA into a smaller space within the bacterial cell. The binding of DNA by H-NS and the resulting changes in DNA structure, can impact the expression of genes by making regions of DNA inaccessible to RNA polymerase.

HU and IHF are homologous proteins that act in a similar way upon DNA

The HU and IHF proteins are homologous to one another, being similar in sequence and activity, yet the IHF protein has sequence specificity in its DNA binding, while the HU protein does not. The binding of IHF to DNA generates a bend in the DNA of 140°−180° at the sites that are specifically recognized by IHF due to its sequence.

HU, with no sequence specificity, instead binds based on the DNA structure. Supercoiled DNA is bound by HU preferentially. HU also bends DNA, like its homologue IHF. HU also binds to structures in the DNA like nicks, gaps, and junctions. HU recognizes the bends that are caused by these structures and stabilizes them.

The Fis nucleoid protein is involved in the regulation of rRNA transcription

Fis binds to DNA and bends it, in bends of 50°−90°. Fis recognizes a binding sequence, but this sequence is poorly conserved and the protein can also bind in a non-specific way to DNA. In *E. coli*, Fis binds to the chromosome at roughly once every 230 bp.

In addition to its structural role within the nucleoid, Fis is involved in the activation of transcription of rRNAs. Ribosomal RNAs are highly transcribed regions of the genome. Fis binds to three sites that are 5′ of the rRNAs and enhances transcription by RNA polymerase.

Quorum sensing causes transcriptional changes within the bacterial cell

For some bacteria, it is energetically expensive to express certain genes and their resulting proteins if there are too few bacteria in a local area. When the bacterial population is large, expression of the genes can be beneficial, often not just to the bacteria expressing these genes, but also to other cells within the population.

To achieve an assessment of the number of bacteria nearby, some bacterial species produce a chemical signal that nearby bacteria are able to sense. If the cell is alone, it will receive little to no signal; the only cell making the signal molecule is itself. If the cell is in a large population with other bacteria like itself, then it will receive a strong signal, indicating that the population size is large. This process is called **quorum sensing**.

Quorum sensing allows the bacterial population to adapt and survive in the environment in a different way when the population is large versus when it is small. The most common and well-studied signaling molecules are **N-acyl homoserine lactones**, but other signal molecules exist as well. The signaling molecule is exported from the cell into the extracellular environment. The bacteria are also able to take up the signaling molecule from the environment and recognize it as a quorum sensing molecule. If the concentration of signal is high, then this corresponds to a high population and the expression of genes is induced (**Figure 5.19**). More detail about how bacteria use quorum sensing is explored in Part V.

Figure 5.19 Quorum sensing enables bacteria to assess the size of its population. Diffusion of a small molecule (red) made by the bacteria can be sensed by the bacteria and used to account for the size of the population. As the bacterial population grows (left to right), the density of signaling molecules increases. Once the population density has reached a key point, the bacteria sense enough of the quorum sensing molecule nearby and alter their gene expression (denoted by change in color).

Biofilm formation is a specialized response to quorum sensing and other signals

Biofilms are formed by microorganisms that have come together as a community with each cell contributing to the whole. The biofilm structure is particularly robust. It adheres to surfaces and due to its make-up is resilient to stresses, including antibiotics. Biofilm formation is complex, requiring the controlled regulation of several genes that must be expressed in the correct order.

Quorum sensing can trigger the formation of a biofilm. It is believed that the bacterial cells are better able to form a biofilm when there are enough of them in one place. The cell can assess this state using quorum sensing. Other signals can also be involved in the bacterial cells' switch from free-living organisms to a biofilm. These include the presence of antibiotics, the presence of a suitable surface on which to build the biofilm, and the availability of nutrients and host-derived cues.

The biofilm includes extracellular sugars as **exopolysaccharide**, therefore, the synthesis and export of polysaccharides must be up-regulated for biofilm formation. There are also extracellular proteins and DNA within the biofilm, as well as the bacteria themselves. The specific biofilm matrix proteins are transcriptionally regulated, as is the DNA export system. DNA can be released from bacterial cells by lysis and in this way contribute to the extracellular DNA. Some bacteria are capable of autolysis upon receipt of certain cues. However, living bacterial cells can contribute DNA as well, through the release of membrane blebs containing DNA. The generation of DNA-containing blebs can also be up-regulated. Bacterial biofilms are explored in more detail in Part V.

Cyclic di-GMP is involved in the regulation of a range of functions within the bacterial cell

The state of the environment that the bacteria cell is in and changes to that environment can be sensed in a number of ways. One is through the creation and degradation of a secondary molecule that acts as a signal for the state of the environment. Two molecules of GTP can come together to form **cyclic di-GMP** through the enzymatic action of **diguanylate cyclases**. Cyclic di-GMP is broken down by **phosphodiesterase** enzymes. Often the activity of these two enzymes is modulated by environmental conditions. The cyclic di-GMP is able to bind to a range of receptors that regulate gene expression, including riboswitches and transcription factors. In this way, cyclic di-GMP is able to regulate diverse cellular functions including biofilm formation, motility, and expression of virulence factors including those involved in adhesion.

The small molecule ppGpp is an indicator of the state of the bacterial cell

Within bacterial cells, the guanosine tetraphosphate (ppGpp) molecule varies in its quantity depending on the state of the cell. High levels of ppGpp accumulate within cells that are starved of amino acids. These cells enter a stringent response where synthesis of RNA is reduced including inhibition of tRNA and ribosome synthesis, as well as production of mRNAs. Some global regulators, including IHF, Lrp, and σ^S, rely on ppGpp for their expression.

Protein thermosensors regulate expression of proteins via transcriptional regulation

There are protein thermosensors, in addition to the RNA thermosensors from Chapter 4. Heat-shock proteins are needed as chaperones to prevent proteins from misfolding or degrading at increased temperatures. Many proteins in the bacterial cell are prone to misfolding at higher than normal temperatures, including regulatory proteins. The misfolding of a regulatory protein will change the structure of the regulator as the temperature increases. At high temperatures, the DNA binding activity of repressor proteins and activator proteins is poor, due to the temperature-induced structural changes in the protein. In the absence of stabilization of the regulator protein structure by heat-shock protein chaperones, the regulator fails to act. Temperature increases can either end the repression of a gene by the repressor protein or prevent the activation of a gene by an activator protein, influencing the transcription of the gene.

Stability and degradation of mRNA by ribonuclease III impacts upon whether a transcript is expressed as a protein

Several criteria can induce mRNA to be degraded by RNases, including the formation of bends in the mRNA and the presence of double-stranded RNA. The bacterial cell contains several different types of RNase enzymes that are able to recognize different features of RNA, including when it is double stranded, when it is folded into a hairpin, when it is occupied by ribosomes, and when a ribosome is stalled at the 3′ end of an mRNA inducing non-stop decay.

RNase III is a global regulator of gene expression. Through its action degrading mRNA, it stops the production of proteins. RNase III binds to double-stranded RNA and cleaves phosphodiester bonds on both strands of segments of double-stranded RNA. Antisense RNAs binding to mRNA form double-stranded RNA that is targeted by RNase III. The secondary structures formed by mRNA folding upon itself generates double-stranded RNA.

Regulation of gene expression can be through the action of RNA binding proteins

The regulation of expression from a transcript can occur through direct action that influences whether the complete transcript is made. Expression can be controlled both during and after transcription of the mRNA through the actions of RNA binding proteins. These proteins can change the susceptibility of an mRNA to degradation by RNases, change the availability of the ribosome binding site (RBS), aid in other cell components interacting with the mRNA, and influence the generation of the complete transcript through encouraging the formation of transcriptional terminators or anti-terminator secondary structures.

Some RNA binding proteins are RNases, which bind to RNA and cleave it, as previously discussed in this chapter. Other RNA binding proteins can block the action of RNases by binding to RNase recognition sequences, blocking them from being bound by RNase. RNA binding proteins can also change the secondary structure of an mRNA, protecting it from RNases that recognize such structures or making the mRNA more susceptible to degradation due to structural changes.

Unless the RBS is available for recognition by the complementary sequence in the 30S ribosomal subunit's 16S rRNA, translation cannot initiate. RNA binding proteins can influence translation by binding to the RBS region or causing the region to form secondary structures. If the mRNA is not covered by ribosomes engaged in translation, it is prone to degradation. RNA binding proteins can stabilize the formation of a transcriptional terminator, which prematurely stops transcriptional elongation, especially of mRNAs that are not being translated. Alternatively, binding of an RNA binding protein can result in an anti-terminator structure forming, which stops the transcriptional terminator structure from forming.

Some RNA binding proteins bind to both mRNA and another molecule in the cell, a small noncoding RNA or protein. In the case of ncRNAs bound by RNA binding proteins, the protein aids in the complementary binding of the ncRNA to the mRNA RBS region.

Key points

- Transcriptional regulation alters the levels of mRNA in the cell for the genes encoded on the DNA.

- The regulation of expression of the co-transcribed *lac* operon genes *lacZYA* is influenced by the presence of glucose and lactose, mediated by the regulatory proteins LacI and CAP-cAMP.

- Transcription of the *trp* operon, encoding genes for the synthesis of tryptophan, is regulated by the presence of tryptophan within the cell, which alters the binding properties of the repressor protein and influences the formation of mRNA secondary structures.

- Regulatory proteins can orchestrate synchronized gene-expression changes across the genome through binding to DNA sites associated with distant genes.

- Two-component regulatory systems sense stimuli through the sensor kinase, which in response autophosphorylates and then passes the phosphoryl group to the response regulator, which regulates gene expression.

- Changes in the DNA sequence associated with the promoter region can influence the transcription initiated from the promoter.

- Alternative sigma factors can coordinate the expression of genes across the genome.

- DNA binding proteins within the nucleoid can have a direct impact on gene expression due to the availability of the DNA for the initiation and elongation of transcription.

- Bacteria can sense the size of the population through quorum sensing and orchestrate the expression of population-dependent characteristics such as biofilm formation.

Terms, questions, and discussions

Key terms

Define the following terms introduced in this chapter. Check your answers using the definitions in the Glossary. These terms are also available as Flashcards on-line.

β-Galactosidase	Chromatin	Phosphorelay
β-Galactoside permease	Constitutive expression	Proteolysis
Anti-sigma factor	Corepressor	Quorum sensing
Anti-terminator	Cyclic AMP	Regulated expression
Attenuation	Effector molecule	Repressor protein
Attenuator	Exopolysaccharide	Response regulator
Autophosphorylate	Methylation	Sensor kinase
Biofilm	*N*-acyl homoserine lactones	Transcriptional regulation
Catabolite activator protein	Operator	Tryptophan
Catabolite repression	Operon	Two-component regulatory system

Self-study questions

Answer each question using 50–100 words or a table or labeled diagram. Advice on where to find answers to these questions is available on-line.

1. What cellular process must occur for a gene to be translated into a protein and how might the bacterial cell alter the levels of production of the protein encoded by a gene?

2. List three situations, not using examples from this chapter, where a bacterial cell may regulate its gene expression to ensure the correct or most efficient proteins are being produced.

3. Draw the genes involved in the *lac* operon locus, including the gene upstream of the *lac* operon itself. Label each of the genes.

4. Add to the diagram from question 3 labels for each of the features involved in the regulation of expression of the *lac* operon.

5. Explain what happens to the transcription of the *lac* operon in *E. coli* when glucose levels are low and lactose is present. Use the diagram you have drawn for reference.

6. Compare your answer to question 5 with what happens at the *lac* operon when glucose levels are high, first when there is no lactose, and then when there is lactose present.

7. Draw the different ways in which the sequence 5′ of the *trp* operon can form hairpins when it is transcribed into mRNA. Label the features of this region.

8. Describe what happens in the regulation of the *trp* operon when levels of tryptophan are low. How does this influence the progression of both the ribosome and RNA polymerase and therefore the transcription of *trpEDCBA*?

9. Contrast your answer to question 8 to what would happen when levels of tryptophan are high. How does the progression of the ribosome and RNA polymerase differ and how therefore does transcription of the *trp* operon differ?

10. What are the key components of a two-component regulatory system and what are the steps that occur in such a system when a signal is detected?

11. How does a phosphorelay compare to a two-component regulatory system? Include features that are similar as well as those that are different.

12. How can sigma factors, a subunit of RNA polymerase, be involved in the control of gene expression?

Discussion topics

These topics are presented for discussion in study groups, as part of class discussions, or on your own. These questions go beyond what is directly covered in this part of the book. Use the research literature and other reading to explore these topics in more depth. Tips to help prepare for topic discussions are available on-line.

13. The *lac* operon and the *trp* operon are often taught as key examples of gene regulation in bacterial systems, however, there are many other regulatory systems in a variety of other bacterial species. Investigate one other bacterial operon and the way(s) in which it is regulated. Explore whether there are repressor proteins, activators proteins, protein binding sites, effector molecules, ncRNAs, riboswitches, and so on.

14. Two-component regulatory systems are often involved in important cellular processes related to host cell interaction or pathogenesis. Investigate a bacterial species of interest and identify a two-component regulatory system or phosphorelay that has been the topic of research. Discuss how the bacteria uses this system to sense its environment and orchestrate the expression of genes needed for the situation in which it finds itself.

15. Discuss the ways in which phase variation is different from gene-expression regulation, although both result in differential expression of genes through varying the transcription of DNA to mRNA.

Further reading

Operons

Jacob F, Monod J. Genetic regulatory mechanisms in the synthesis of proteins. *J Mol Biol.* 1961; 3 (3): 318–56.

Ralston A. Operons and prokaryotic gene regulation. *Nat Educ.* 2008; 1(1): 216.

Chromatin and regulation

Dame RT. The role of nucleoid-associated proteins in the organization and compaction of bacterial chromatin. *Mol Microbiol.* 2005 May; 56(4): 858–70.

Wade JT, Grainger DC. Waking the neighbours: Disruption of H-NS repression by overlapping transcription. *Mol Microbiol.* 2018 May; 108(3):221–5.

Other forms of expression regulation

Tiwari S, Jamal SB, Hassan SS, Carvalho PVSD, Almeida S, Barh D, Ghosh P, Silva A, Castro TLP, Azevedo V. Two-component signal transduction systems of pathogenic bacteria as targets for antimicrobial therapy: An overview. *Front Microbiol.* 2017; 8: 1878.

Tucker BJ, Breaker RR. Riboswitches as versatile gene control elements. *Curr Opin Struct Biol.* 2005; 15(3): 342–8.

Zelewska MA, Pulijala M, Spencer-Smith R, Mahmood HA, Norman B, Churchward CP, Calder A, Snyder LA. Phase variable DNA repeats in *Neisseria gonorrhoeae* influence transcription, translation, and protein sequence variation. *Microb Genom.* 2016; 2(8):e000078.

Transcriptomes

The term transcriptome refers to all of the RNA transcripts that are present within the cell. In some instances, transcriptome may be used to refer to just the mRNAs within the cell, representing the genes that are being transcribed and thus likely to be expressed as proteins. Transcriptome can also include other RNAs within the bacterial cell, such as the ncRNAs that influence the expression of genes. Also transcribed in the cell, and can therefore be encompassed in the term transcriptome, are the rRNAs and the tRNAs; however, the transcription of these tends to be more constitutive, while the transcription of mRNAs and ncRNAs can vary depending on the conditions impacting the cell. Regardless of the breadth of RNA types included, it is the variations in the amounts of the different distinct transcripts within the cell that is the focus of transcriptomics.

Only a portion of the genes in the genome are expressed at any one time. As discussed in Chapter 5, the regulation of genes means that the generation of some of the transcripts within the bacterial cell is controlled. As a consequence, the RNAs that constitute the transcriptome are not constant. The levels of rRNA and tRNA present within the bacterial cell are relatively constant; if they do change, these changes are not specific to a particular gene product. Changes in rRNA and tRNA may indicate an increase or decrease in general protein production, rather than being specific for any particular products. The mRNAs and ncRNAs are linked to specific gene products; therefore, increases or decreases in particular mRNA or ncRNA transcripts over time are expected within the transcriptome. Studying the transcriptome can provide insight into how bacterial cells use their genes and regulate the expression of their genes in changing environments.

The transcriptome changes over time

The transcriptome changes over time and in response to conditions impacting the bacterial cell. Changes to the transcriptome can occur as regulated genes are switched on and off, being up- and down-regulated in their expression. As seen in Chapter 5, the presence of lactose in the absence of glucose will cause the transcription of the *lac* operon. However, when glucose is present, the *lac* operon is not transcribed. Therefore, the presence in the transcriptome of the *lac* operon mRNA is not constant and is influenced by the presence of lactose and glucose within the bacterial cell. As the bacteria transition from stimulus to stimulus, they adapt to these changes and make best use of their resources by differentially expressing sets of genes (**Figure 6.1**). The transcriptome therefore changes over time as a result of the stimuli to which the bacterial cell is exposed and the corresponding activation and repression of transcription.

Changes to the transcriptome over time can also occur due to mutation, which changes the genomic template for transcription. Environmental pressures may

Figure 6.1 The transcriptome changes over time. As bacterial cells encounter various stimuli, the sets of genes that are expressed changes. In response to a stimulus (arrow 1), the expression of some genes will be activated (dotted line) and others repressed (dashed line). When the cell encounters a different stimulus (arrow 2), a different set of genes is turned on (dashed line) and others turned off (dotted line). Thus, the RNAs that make up the transcriptome change over time. Some RNAs that make up the transcriptome remain relatively unchanged, including the mRNAs for housekeeping genes, rRNAs, and tRNAs (solid line).

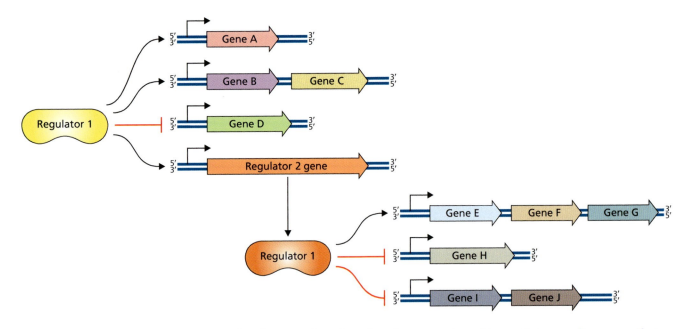

Figure 6.2 Regulatory networks alter the transcription of several genes. When a regulatory protein or regulatory ncRNA (regulator 1) is involved in the regulation of several genes, the activation (arrows) and repression (red line) of the genes within this network are coordinated together. Regulators have the capacity to regulate other regulators (regulator 2), thus generating more complex regulatory networks.

select for mutations that alter the expression of transcripts encoding genes that are beneficial to bacterial growth in such an environment. These sorts of adaptations can occur as a result of the presence of antibiotics or during the process of host adaptation or adaptation to growth in a laboratory. A mutation that causes the loss of a regulatory protein or regulatory ncRNA can affect the presence of transcripts within the transcriptome that are created as a result of these regulators (**Figure 6.2**). Changes to a regulator may influence the expression of just one gene or operon, or it may impact several genes if it is part of a **regulatory network**. Genomic changes such as sequence inversions, duplications, or deletions can alter the regulation and transcription of features associated with the affected region, changing the transcriptome. Mutations in the genome at the DNA level therefore have a knock-on effect to the levels of the various RNAs within the bacterial cell transcriptome.

Changes in the transcriptome can arise either through transcriptional control or through genetic mutation, with differing consequences for each. While changes over time due to changes in conditions encountered by the bacterial cell are transient, changes to the genome sequence that impact the transcriptome can be permanent. It is therefore important to put transcriptomic discoveries into the context of the genome. While the presence or absence of the *lac* operon mRNA in the transcriptome may be due to the presence of lactose and glucose, it may also be due to a mutation resulting in the loss of LacI or a change in the sequence of the operon promoter that changes the strength of that promoter for recruiting RNA polymerase. When assessing the transcriptome, it is important to consider the genome as well, to determine if changes are due to regulation or mutation.

The transcriptome changes due to changing conditions

The growth rate of bacterial cells is influenced by the synthesis of the ribosomes that are required for translation, enabling the cell to produce the proteins needed for growth. Ribosomes are made up of both rRNAs and also ribosomal proteins. The expression of ribosomal proteins requires the transcription of their genes into mRNA by RNA polymerase and subsequent translation by existing ribosomes. Transcription of the rRNAs is also required to form ribosomes (**Figure 6.3**). When the bacterial cell experiences stress, there is a decrease in the growth rate. The regulation of gene expression coordinates the growth rate of the bacterial population through orchestration of the levels of ribosomes, of the production of new cellular components needed for cell division, and of the cellular processes such as genome replication.

Responses to a variety of environmental cues alter gene expression within the bacterial cell and these changes in the transcriptome dictate the rate at which

Figure 6.3 The formation of ribosomes from proteins and RNA. Ribosomes are required for translation of mRNAs into proteins. New ribosomes are synthesized within the bacterial cell through the combination of ribosomal proteins and ribosomal RNA (rRNA). The genes encoding ribosomal proteins are transcribed by RNA polymerase into mRNA and translated by ribosomes into ribosomal proteins. There are approximately 32 ribosomal proteins, depending on the species, in the 50S subunit, encoded by genes in various locations in the chromosome. There are approximately 22 ribosomal proteins in the 30S subunit of the ribosome. The genomic loci encoding the rRNAs are also transcribed by RNA polymerase. There is often more than one rRNA locus in a bacterial genome, each with 16S rRNA, 23S rRNA, and 5S rRNA loci. The rRNA and protein products form together into ribosomes.

bacteria can viably divide and the population expand. It was recognized by Torbjörn Caspersson, Jean Brachet, and Jack Schultz in 1939, that cells engaged in making an abundance of proteins, such as rapidly growing bacteria, were rich in RNA. By implication, this finding suggested that RNA is required for cells to make proteins, which was later shown to be due to the process of transcription. This early observation about the nature of RNA and differences in the levels of RNA within bacterial cells, gave insight into how the transcriptome can change due to changes in cellular processes, just as studying transcriptomes can today.

In a growing population of bacterial cells, a variety of stresses may be encountered and must be overcome to ensure continued survival of the population. If a needed sugar or amino acid is scarce, the bacterial cell must compensate for this shortage of nutrients in order to survive. Each stress triggers the regulation of an associated cascade of gene expression, altering the overall transcriptome in response to the stress. For this reason, transcriptome studies can be complex. Slight changes in the environment can cause changes in the transcriptome, therefore careful control of experimental conditions and replication of experiments are key to producing informative results.

Expression of genes outside the *lac* operon in response to glucose and lactose

The presence of glucose and lactose within the cell dictates the expression of the genes in the *lac* operon, as discussed in Chapter 5. At a transcriptomic level, genes outside of the *lac* operon also vary in their expression level in response to the presence of these sugars. CAP–cAMP positively regulates the *lac* operon in the absence of glucose, but the *lac* operon is not the only feature regulated by CAP. The gene for CAP is not part of the *lac* operon; the gene is found elsewhere on the chromosome. In conjunction with cAMP, CAP activates several genes within the *Escherichia coli* genome, including those for sugar metabolism such as the *lac* operon, but also other genes that are expressed in response to low glucose (**Figure 6.4**). CAP–cAMP is also able to repress the expression of genes, therefore it is also referred to as **CRP, cAMP receptor protein**.

Evolutionary theory suggests that cellular processes will be stopped if they are not needed, in order to conserve and best use cellular energy. One theory

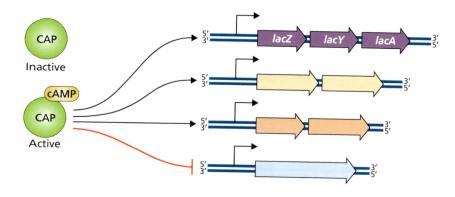

Figure 6.4 CAP–cAMP regulates the *lac* operon and also other genes in the *E. coli* genome. The regulatory protein CAP is inactive and does not regulate gene expression when it is alone. CAP becomes active when it is complexed with the metabolite cAMP, which is present in the bacterial cell when glucose is low. The CAP–cAMP complex activates the genes of the *lac* operon (arrow) and also regulates other genes in the *E. coli* genome (activation, arrow; repression, red line).

Figure 7.3 The partial double bond character of a peptide bond. Electron pairs (dots) from the amine end of one amino acid contribute to the peptide bond between the amine end of one amino acid and carboxyl end of another. This makes the nitrogen positively charged and the oxygen negatively charged, contributing to the strength of the covalent bond between the carbon and nitrogen.

The inflexible nature of the peptide bond imposes limits on the amino acids

Peptide bonds are slightly stronger and slightly shorter than other covalent single bonds and therefore have more of a double bond character. This is due to the presence of a lone pair of electrons on the nitrogen that contribute to the peptide bond (**Figure 7.3**).

Due to the double bond-like nature of the peptide bond, there is reduced rotation between two joined amino acids than there would be with a single bond. The peptide bond tends to be in a ***trans*** configuration, with side chains of adjacent amino acids on opposite sides from one another (**Figure 7.4**). This formation is favored due to **steric hindrance** where the amino acid side chains interfere with one another structurally and this interference cannot be overcome by rotation due to the inflexible peptide bond. The rigid nature of the peptide bond therefore tends to have amino acids in *trans*, however, proline tends to be in a ***cis*** configuration due to its odd structure (described later in this chapter), where the side chain connects back to the N of the amine group.

Rotation within the peptide chain is therefore associated with the α carbon, which retains its flexible single bond. The juxtaposition of the flexible bonds at the α carbon and the relatively inflexible peptide bond between the amino acids influences the protein structures that can form. Peptide chains tend to be drawn in *trans*, with the N-terminus on the left and C-terminus on the right (**Figure 7.5**).

Amino acids are generally present as zwitterions

The carboxyl group of an amino acid is a weak acid. It can lose its hydrogen to become $-COO^-$. The amino acid's amine group, in contrast, is a weak base. It can gain a hydrogen to become $-NH_3^+$. In pH range 2.2–9.4, which includes general physiological pH conditions for most bacteria, an amino acid is generally present with a $-COO^-$ end and an $-NH_3^+$ end. This is called a **zwitterion**, having a negatively charged end and a positively charged end, and a net charge of zero. The zwitterionic form is the usual state of an amino acid or peptide, although they are frequently shown with –COOH and –NH2 ends (**Figure 7.6**).

trans   cis

Figure 7.4 *Trans* and *cis* conformations of amino acids. Due to the double bond character of the peptide bond between amino acids, the side chains are locked into one side or the other of the molecule. When two adjacent amino acids have side chains on opposite sides this is *trans* (left). When both side chains are on the same side, this is *cis* (right).

Figure 7.5 Typical illustration of a peptide. The N-terminus is on the left and the C-terminus is on the right. The side chains are on opposite sides in the *trans* configuration, as is typical for most amino acids. Rotation around the α carbon (red) is more flexible than the peptide bond (blue).

Figure 7.6 Zwitterionic form of a peptide. At normal pH, the carboxyl group will tend to lose a hydrogen and the amine group will tend to gain one. This gives the molecule a positive charge at the amine end and a negative charge at the carboxyl end, but a zero net charge overall. Generally, peptides are zwitterionic, with a positive charge on the N-terminus and a negative charge on the C-terminus.

There are 20 amino acids encoded in the standard genetic code of DNA

There are 20 standard amino acids that are encoded by codons in the DNA and carried by mRNA to be translated at the ribosome. Amino acids that can be found in proteins are called **proteinogenic amino acids**. Although hundreds of different amino acids exist, most are not encoded in DNA and therefore are not found in proteins. The specific amino acids within a protein are determined by the gene encoding it in its DNA. The codon sequences of three nucleotides each correspond to one of the 20 amino acids or are a signal to terminate translation.

The four nucleotides within DNA, adenine, thymine, guanine, and cytosine, can come together in 64 different three base codon combinations in the mRNA as adenine, uracil, guanine, and cytosine. Three of these 64 combinations are the stop codons that terminate translation. The remaining 61 codons correspond to amino acids, but these 61 codons encode only 20 amino acids, which means that most amino acids have more than one codon. The exception is methionine, which has only the codon AUG, and tryptophan, which has only the codon UGG. Arginine, leucine, and serine have six codons, but most amino acids have two codons (see **Table 2.1**).

Most of the amino acids have more than one codon. This means that if a mutation in the nucleic acid sequence occurs, the new codon has the potential to still correspond to the same amino acid. For example, the codon GGU changing to the codon GGC due to a mutation that affects the last base of the codon, does not change the encoded amino acid, as both of these are codons for glycine. If the nucleic acid sequence changes the codon to a different amino acid, that amino acid might still have similar characteristics to the original amino acid. For example, if the codon CUU encoding the amino acid leucine changes to the codon AUU encoding isoleucine, the resulting protein may be unaffected due to the similarity of these two amino acids. However, a hydrophilic amino acid could be changed into a hydrophobic amino acid based on the change in DNA and mRNA sequence, or it could be changed from an amino acid with an uncharged side chain to a charged one. Occasionally there is translational read through of stop codons, where mutated tRNAs may recognize a termination codon as an amino acid encoding codon. For example, UAA terminator is interpreted as CAA for Gln, UAU or UAC for Tyr, or AAA for Lys due to a mutation in the tRNA anticodon. A single base change in a tRNA could result in the stop codon UAG being recognized as CAG for Gln, UAC or UAU for Tyr, or AAG for Lys.

The codons present in the DNA sequence is species specific

It is possible to predict the amino acid sequence of a protein from the DNA sequence. The codons and their corresponding amino acids are known and therefore the ability to predict the protein sequence from a DNA sequence is robust. However, it is not possible to reverse such predictions and to accurately predict the DNA sequence from a protein amino acid sequence. Predicting a DNA sequence from a protein sequence is not straightforward because some amino acids have six codons, some have four codons, and most have at least two. Because there are options for different codons that will encode the same amino acid, variations in the use of codons are seen between the bacterial species.

Some codons are preferred over others, dependent upon the bacterial species. Therefore, although most amino acids can be encoded by more than one codon, there will be one that is used most often in the coding sequences of bacterial genes based on the species. If a codon is rare or used infrequently in a genome, there is likely to be less of the corresponding tRNA. The use of rare codons with few tRNAs within the cell results in decreased stability for mRNAs containing the rare codon. Slow or stalled translation as a result of the scarcity of the tRNA will mean that the mRNA is targeted for degradation. Because codon usage is species specific, analysis of the codon usage within regions of the genome can identify regions of DNA where codon usage is different from the rest of the genome. Regions with different codon usage can indicate sections of the bacterial genome that have been acquired from another species.

The prevalence of the different termination codons, necessary for the proper ending of an amino acid chain, can be influenced by the species-specific G+C content of the genome. For instance, the frequency of TGA as a stop codon generally increases with the G+C content of the genome, whereas the stop codon TAA increases with lower genomic G+C. However, the G+C influence on stop codon usage generally does not apply to the termination codon at the end of a gene. The bias applies only to termination codons present in the other five reading frames, those that are not used by a coding sequence and therefore prevent erroneous proteins from being made from noncoding frames of the DNA. Stop codons are plentiful in the reading frames that are not used by the coding sequence and it is here that there will tend to be TGA in high G+C genomes and TAA in low G+C genomes.

The amino acids that can be made by bacteria are species specific

Some amino acids can be essential for the bacterial cell because the amino acid cannot be made by the cell from other components. The cell must therefore acquire the amino acid as it would other nutrients for growth. The amino acids that are essential, those that cannot be made by the cell, differ depending on the species of bacteria. The requirement for essential amino acids in the growth media can help differentiate and identify bacterial species.

Most amino acids are L stereoisomer α-amino acids

The carbons in amino acids are numbered starting from the –COOH group. The second carbon along from the carboxyl group is called the α-carbon and this is where the side chain of the amino acid is found. Most amino acids found in nature have the –NH$_2$ group bonded to the α-carbon, making them α-**amino acids**.

Amino acids can occur in two different structures that are mirror images of each other (**Figure 7.7**). Glycine is the only amino acid that does not have stereoisomers. The structure of glycine has two hydrogens attached to its α-carbon; the R side chain is simply another hydrogen. Therefore, the mirror images are identical, so it does not have different stereoisomers. Most amino acids are L **stereoisomers**, referring to the three-dimensional structure of the amino acid and the position of the side chain. L amino acids are incorporated into proteins during translation. D amino acids, the mirror image of L amino acids, can be added to the protein during **post-translational modification**, a process where changes are made to a peptide after translation. D amino acids are also part of the bacterial peptidoglycan cell wall on the outside of Gram-positive bacteria and between the two membranes of Gram-negative bacteria.

Figure 7.7 L and D stereoisomers of the amino acid alanine. In alanine the side chain attached to the α-carbon is –CH$_3$. Stereoisomers of alanine will have the –CH$_3$ side chains as mirror images in the structure, with the L stereoisomer having the side chain on the left (left) and the D stereoisomer having the side chain on the right (right).

The classification of an amino acid is determined by its side chain

The side chains of amino acids are added onto the α-carbon and any subsequent carbons in the structure are therefore designated by a Greek letter: β; γ; δ; and so on (**Figure 7.8**). Some amino acid side chains are linear but not all of them. **Branched chain amino acids** that have side chains with branches in their chemical structure include leucine, isoleucine, and valine.

Amino acids are classified based on the properties possessed by the side chain, which include being weak acids or weak bases, being hydrophilic or hydrophobic, and being polar or nonpolar. These properties of the amino acids within the protein determine the function of the protein, the structure of the protein and how it folds, and the stability of the structure of the protein.

Figure 7.8 Designation of carbons in an amino acid. The carbon to which the amino acid side chain is attached is referred to as the α-carbon. Subsequent carbons attached to the α-carbon in the side chain are designated by Greek letters.

Hydrophobic side chains tend to be buried in the middle of proteins that are found in the bacterial cytoplasm. Hydrophobic amino acids are also found in proteins that are within the bacterial membrane, which is itself hydrophobic.

Proteins that act by binding negatively charged molecules, such as DNA, have positively charged amino acids on their surface. Therefore, the positively charged surface of a regulatory protein may aid in the binding of the protein to the operator, thereby influencing transcription from a promoter.

The features and classifications of amino acids contribute together to the overall features of proteins containing them. The structure of the protein is influenced by factors such as the inflexible nature of the peptide bond and whether that peptide bond is *trans* or *cis*. The potential for hydrogen bonding between the oxygen on the carboxyl group and the hydrogen on the amine groups of different amino acids within a peptide is influenced by the flexibility of the peptide chain and any steric crowding caused by the nature of the side chains, the charge of the side chains, and the hydrophobic or hydrophilic nature of the side chains.

Glycine is small and flexible

The most flexible of all of the amino acids structurally is glycine. It is able to easily rotate because the side chain is simply a hydrogen (**Figure 7.9**), therefore it does not generate steric hindrance that might limit its ability to rotate. Structures including glycine are not restricted by interference from its side chain, unlike other amino acids, and therefore it can be found both inside and outside of a protein. With such a small side chain, glycine is the smallest of the amino acids. Glycine is abbreviated as Gly or G and is encoded by four codons: GGU; GGC; GGA; and GGG.

Figure 7.9 The structure of the amino acid glycine. Glycine is the smallest and most flexible of the amino acids, with a hydrogen for its side chain.

Alanine is abundant and versatile

Alanine is small like glycine, but is less flexible than glycine. It has –CH_3 as a side chain (**Figure 7.10**), which is hydrophobic and polar. Like glycine, the flexible nature of the alanine amino acid means that it is versatile. Alanine is also abundant in the bacterial cell, therefore it can be found in many proteins. Alanine is abbreviated as Ala or A and is encoded by four codons: GCU; GCC; GCA; and GCG.

Figure 7.10 The structure of the amino acid alanine. Alanine has a hydrophobic side chain containing a carbon and three hydrogens.

Arginine has a positively charged side chain

The side chain of the amino acid arginine is positively charged, containing carbons, nitrogens, and hydrogens (**Figure 7.11**). Arginine is often found in proteins that bind to phosphorylated substrates. Due to similarities in the features of the side chain, arginine and lysine (see Figure 7.21) are similar. Arginine is abbreviated as Arg or R and is encoded by six codons: CGU; CGC; CGA; CGG; AGA; and AGG.

Asparagine was the first amino acid identified

The first amino acid identified was asparagine, isolated from asparagus by Louis-Nicolas Vauquelin and Pierre Jean Robiquet in 1806. It is acidic, with an uncharged side chain (**Figure 7.12**), and is similar to aspartic acid. The amine within the

Figure 7.11 The structure of the amino acid arginine. Arginine has a positively charged side chain containing four carbons, three nitrogens, and 11 hydrogens.

Figure 7.12 The structure of the amino acid asparagine. Asparagine has an uncharged side chain containing two carbons, a nitrogen, an oxygen, and four hydrogens.

Figure 7.13 The structure of the amino acid aspartic acid. Aspartic acid has a negatively charged side chain containing two carbons, two oxygens, and two hydrogens.

asparagine side chain is easily hydrolyzed, which results in asparagine changing to aspartic acid. Asparagine is involved in hydrogen bonding and is a common site to attach carbohydrates in glycoproteins. Asparagine is abbreviated as Asn or N and is encoded by two codons: AAU and AAC.

Aspartic acid is negatively charged and binds to positively charged molecules

The side chain of aspartic acid is negatively charged, containing carbons, oxygens, and hydrogens and is acidic (**Figure 7.13**). This charge means that aspartic acid is able to bind to positively charged molecules and as such is one of two acidic amino acids involved in enzymatic active centers. It is both acidic and hydrophilic, often being found on the outer surface of proteins. Aspartic acid is abbreviated as Asp or D and is encoded by two codons: GAU and GAC.

Figure 7.14 The structure of the amino acid cysteine. Cysteine has a side chain containing a carbon, a sulfide, and three hydrogens.

Cysteine forms disulfide bonds with other cysteines

The amino acid cysteine has sulfide in its side chain (**Figure 7.14**), which can form covalent **disulfide bonds** to other cysteines. The sulfides that are within the side chains of the cysteines come together and the resulting disulfide bonds stabilize the protein structure. The sulfur within the side chain of cysteine can also bond readily to heavy metal ions. Cysteine is abbreviated as Cys or C and is encoded by two codons: UGU and UGC.

Figure 7.15 The structure of the amino acid glutamic acid. Glutamic acid has a negatively charged side chain containing three carbons, two oxygens, and four hydrogens.

Glutamic acid is a large, acidic amino acid

Glutamic acid is similar to aspartic acid in that it is an acidic amino acid with a negatively charged side chain (**Figure 7.15**). However, glutamic acid is large with a longer side chain than aspartic acid and a side chain that is slightly more flexible. Glutamic acid is abbreviated as Glu or E and is encoded by two codons: GAA and GAG.

Glutamic acid can be easily converted to proline. This is accomplished in a three-step process in which glutamic acid becomes glutamate semialdehyde, which becomes Δ'-pyrroline-5-carboxylate, which then becomes proline (**Figure 7.16**).

Glutamic acid   Glutamate semialdehyde Δ'-Pyrroline 5-carboxylate   Proline

Figure 7.16 The conversion of glutamic acid to proline. First, glutamic acid is converted to glutamate semialdehyde. Loss of water from this intermediate produces Δ'-pyrroline-5-carboxylate. This then becomes the amino acid proline.

Glutamine has an uncharged side chain

Glutamine is similar to glutamic acid in its side chain, being acidic and large (**Figure 7.17**). However, glutamine has an uncharged side chain rather than the negative charge of glutamic acid. Glutamine is used to store ammonia in proteins. Glutamine is abbreviated as Gln or Q and is encoded by two amino acids: CAA and CAG.

Histidine has a large positively charged side chain containing a ring structure

The side chain of histidine includes a ring structure made of carbons and nitrogens (**Figure 7.18**). This gives it a large, positively charged side chain. Histidine is commonly found in enzymes and can stabilize the structure of a protein. In acidic conditions, protonation of the nitrogen within histidine occurs, which causes a conformational change in the protein. Histidine is abbreviated as His or H and is encoded by two codons: CAU and CAC.

Isoleucine has a branched side chain

Isoleucine has a branched side chain (**Figure 7.19**) and as such is usually interchangeable with leucine and sometimes valine. The branched side chain of isoleucine is large and hydrophobic. Due to its properties, isoleucine contributes to correct folding and protein stability and is important for ligand binding. Isoleucine is abbreviated as Ile or I and is encoded by three codons: AUU; AUC; and AUA.

Leucine is similar to isoleucine, although its branched side chain is configured differently

Leucine has the same number of carbons and hydrogens in its side chain as isoleucine, although it is configured differently (**Figure 7.20**). With large branched side chains, leucine is similar to isoleucine and valine. Generally, leucine is found buried within folded proteins due to the hydrophobic nature of its side chain. Leucine is abbreviated as Leu or L and is encoded by six codons: UUA; UUG; CUU; CUC; CUA; and CUG.

Lysine has a long, flexible side chain

Lysine has a long side chain made of four carbons, a nitrogen, and 11 hydrogens (**Figure 7.21**), which is flexible along its length. The side chain is also positively charged and as a result interacts with negatively charged molecules within the bacterial cell, such as DNA. Lysine is often found on the surface of hydrophilic

Figure 7.17 The structure of the amino acid glutamine. Glutamine has an uncharged side chain containing three carbons, a nitrogen, an oxygen, and six hydrogens.

Figure 7.18 The structure of the amino acid histidine. Histidine has a positively charged side chain containing four carbons, two nitrogens, and six hydrogens.

Figure 7.19 The structure of the amino acid isoleucine. Isoleucine has a hydrophobic side chain containing four carbons and nine hydrogens in a branched structure.

Figure 7.20 The structure of the amino acid leucine. Leucine has a hydrophobic side chain containing four carbons and nine hydrogens.

Figure 7.21 The structure of the amino acid lysine. Lysine has a positively charged side chain containing four carbons, a nitrogen, and 11 hydrogens.

Figure 7.22 The structure of the amino acid methionine. Methionine has a hydrophobic side chain containing three carbons, a sulfide, and seven hydrogens.

Figure 7.23 The structure of *N*-formylmethionine, a modified methionine. A formyl group (COH) is added to the amine group of the methionine, at the left of the structure. The carboxyl group is then bound to the tRNA$_f^{Met}$, at the right of the structure.

Figure 7.24 The structure of the amino acid phenylalanine. Phenylalanine has a hydrophobic side chain containing seven carbons, six of which are in a ring structure, and seven hydrogens.

proteins, such as those found in the cytoplasm. The features of lysine mean that it is also found to be active in enzymes. Lysine is abbreviated as Lys or K and is encoded by two codons: AAA and AAG.

Methionine is at the start of all translation

Methionine (**Figure 7.22**) has a large hydrophobic side chain and as a result, it is frequently found buried in proteins. It contains a sulfur, like cysteine, but with a methyl group that may be activated. Methionine is abbreviated as Met or M and is encoded by a single codon, AUG.

The first amino acid of any new peptide chain generated by translation is a modified form of methionine, ***N*-formylmethionine (fMet)** (**Figure 7.23**). The start codon of a gene always codes for fMet in bacteria. AUG is the most common initiation codon and encodes methionine. However, the alternative initiation codons, GUG and UUG, still encode fMet at the start of translation and it is fMet that is incorporated at the beginning of the polypeptide chain. There are different tRNAs that start translation and these are loaded with fMet. The genes of the *lac* operon, for example, use the alternative initiation codons GUG for *lacI* and UUG for *lacA*. Sometimes the *N*-formylmethionine is removed after translation and is not part of the final protein sequence.

Phenylalanine has a rigid ring structure side chain

The side chain of phenylalanine is hydrophobic and includes a rigid aromatic group that forms a ring structure (**Figure 7.24**). These ring structures can stack, with the phenylalanine side chains forming a stable structure. Phenylalanine is one of the largest proteinogenic amino acids, along with tyrosine and tryptophan. Due to its hydrophobic nature, phenylalanine is almost always buried within the protein. Phenylalanine is abbreviated as Phe or F and is encoded by two amino acids: UUU and UUC.

The side chain for proline loops back to the amine group

Proline has a side chain that forms a ring structure, however, unlike other proteinogenic amino acids, the side chain ring for proline includes the amine group (**Figure 7.25**). Proline is the only cyclic amino acid. This cyclic form fixes the structure of the amino acid, which is unable to be flexible at the α carbon. The presence of proline causes a bend in the polypeptide chain; this can disrupt protein folding, making kinks in the structure. As a result of its structure, proline tends to be in the *cis* configuration when bound to other amino acids. The side chain is polar and due to its structure is actually classified as an imino acid. In a peptide bond, the proline amino group has no H, therefore it cannot donate H to stabilize protein structure. Proline is abbreviated as Pro or P and is encoded by four codons: CCU; CCC; CCA; and CCG.

Figure 7.25 The structure of the amino acid proline. Proline has a side chain that loops back to the amine group (top). The proline amino acid (red) tends to be in a *cis* configuration with other amino acids (blue) due to its structure (bottom).

Serine has an uncharged side chain that readily donates hydrogen

The side chain of serine has a hydroxyl group (**Figure 7.26**) and the hydrogen from this side chain feature is readily donated. The side chain is uncharged and polar. The features of serine mean that it is often found on the outside of hydrophilic proteins. Serine is abbreviated as Ser or S and is encoded by six codons: UCU; UCC; UCA; UCG; AGU; and AGC.

Figure 7.26 The structure of the amino acid serine. Serine has an uncharged side chain containing one carbon, one oxygen, and three hydrogens.

Threonine is similar to serine with an uncharged polar side chain

The side chain of threonine has two carbons, one oxygen, and four hydrogens (**Figure 7.27**). These form an uncharged side chain that is polar and hydrophilic. Side chain feature similarities mean that threonine is therefore similar to serine. Threonine is abbreviated as Thr or T and is encoded by four codons: ACU; ACC; ACA; and AGG.

Figure 7.27 The structure of the amino acid threonine. Threonine has an uncharged side chain containing two carbons, one oxygen, and four hydrogens.

Tryptophan has a large side chain with a double ring structure

The hydrophobic side chain of tryptophan is large, incorporating a double ring structure (**Figure 7.28**). Tryptophan is the largest amino acid and is similar to phenylalanine and tyrosine. Tryptophan is abbreviated as Trp or W and is encoded by just one codon, UGG.

Figure 7.28 The structure of the amino acid tryptophan. Tryptophan has a hydrophobic side chain containing nine carbons, one nitrogen, and eight hydrogens forming a double ring structure.

Tyrosine has a hydrophobic ring structure side chain

The side chain of tyrosine is large and hydrophobic. It includes a ring structure (**Figure 7.29**), making it similar to phenylalanine and tryptophan. Tyrosine is abbreviated as Tyr or Y and is encoded by two codons: UAU and UAC.

Figure 7.29 The structure of the amino acid tyrosine. Tyrosine has a hydrophobic side chain containing seven carbons, one oxygen, and seven hydrogens.

Valine has a branched hydrophobic side chain, similar to isoleucine and leucine

Valine is one of the few proteinogenic amino acids with a branched side chain, the others being isoleucine and leucine. The side chain has three carbons and seven hydrogens (**Figure 7.30**). Due to its hydrophobic nature, valine is usually found inside of proteins. Valine is abbreviated as Val or V and is encoded by four codons: GUU; GUC; GUA; and GUG.

Bacterial proteins can include other amino acids beyond the 20 with codons

In bacteria, in addition to these 20 amino acids, for which there are codons, there are other amino acids that can be incorporated into proteins during translation or added post-translationally. During translation, some bacteria can incorporate the amino acid **selenocysteine** and a few are able to incorporate **pyrrolysine.** These 21st and 22nd amino acids are both encoded by codons that are otherwise termination codons, signaling the end of translation. However, these termination codons become amino acid codons by nature of other features in the mRNA sequence that trigger the incorporation of selenocysteine and pyrrolysine, rather than inducing translational termination.

Selenocysteine is encoded by the UGA termination codon and the selenocysteine insertion sequence (SECIS) of 60 nucleotides. The presence

Figure 7.30 The structure of the amino acid valine. Valine has a hydrophobic side chain containing three carbons and seven hydrogens.

Figure 7.31 The structure of the amino acid selenocysteine. Selenocysteine has a side chain containing one carbon, one selenium, and three hydrogen.

Figure 7.32 The structure of the amino acid pyrrolysine. Pyrrolysine has a side chain containing 10 carbons, two nitrogen, one oxygen, and 17 hydrogens.

of the SECIS element in the mRNA results in the UGA codon recruiting a selenocysteine tRNA rather than triggering translational termination. Just after the UGA on the mRNA, SECIS makes a stem-loop and this causes the UGA codon to be recognized as the selenocysteine codon, rather than the stop codon. Selenocysteine codons can be identified in the genome sequence data by the conserved bases and secondary structure of the SECIS. Selenocysteine is similar to cysteine, although selenium is in the place of sulfur (**Figure 7.31**), and is abbreviated as Sec or U.

Pyrrolysine is encoded by the termination codon UAG, associated with the PYLIS sequence on the mRNA. Pyrrolysine has a long side chain of 10 carbons (**Figure 7.32**) and is abbreviated as Pyl or O.

In addition, amino acids that do not occur naturally in bacterial proteins can be introduced artificially. **Alloproteins** are engineered proteins that do not occur naturally and that can include non-proteinogenic amino acids.

Key points

- Amino acids come together to make proteins.
- Amino acids have an amine group, a carboxyl group, and a side chain.
- The different side chains have different features that influence the structure and function of the amino acid within the protein.
- The peptide bond is double-bond like, which restricts the flexibility of the peptide chain at this bond.
- There are 20 amino acids encoded by the 61 standard three base codons.
- A 21st and 22nd amino acid are encoded by what is otherwise a stop codon, with the added influence of another sequence within the mRNA.
- Most amino acids have more than one codon.
- The usage of particular codons is species specific.
- Charge, hydrophobicity, and polarity are features of amino acid side chains that influence the function and structure of the protein.

Terms, questions, and discussions

Key terms

Define the following terms introduced in this chapter. Check your answers using the definitions in the Glossary. These terms are also available as Flashcards on-line.

α-Amino acids	Disulfide bond	Secondary amino acid structure
β barrel	fMet	Selenocysteine
Amine group	N-formylmethionine	Stereoisomers
Branched chain amino acids	Post-translational modification	Steric hindrance
Carboxyl group	Primary amino acid structure	Tertiary amino acid structure
Chaperone	Proteome	Zwitterion

Self-study questions

Answer each question using 50–100 words or a table or labeled diagram. Advice on where to find answers to these questions is available on-line.

1. What are the units that make up proteins? How many different types of these units are there and how does this compare to the units that make up DNA?

2. Draw the general structure common to all amino acids. Label the amine group and the carboxyl group.

3. Explain why the peptide bond that joins amino acids together is considered to be inflexible compared to the other covalent bonds within the amino acids.

4. Draw a *trans* dipeptide made up of one cysteine and one isoleucine.

5. Draw a *cis* dipeptide made up of one serine and one alanine.

6. What is the zwitterionic state of a peptide and why is this the usual state in physiological conditions?

7. If there are hundreds of different amino acids, why are only some referred to as proteinogenic amino acids?

8. Which amino acids form covalent disulfide bonds?

9. What is the unusual feature of the structure of proline and how does this influence the structure of proteins containing proline?

10. Which amino acid can be easily converted into another amino acid? Draw the conversion, including the intermediate stages in this three-step process.

11. What are the 21st and 22nd amino acids used by bacteria and how do the codons for these work?

12. How might amino acids with branched side chains and/or ring structures contribute to steric hindrance?

Discussion topics

These topics are presented for discussion in study groups, as part of class discussions, or on your own. These questions go beyond what is directly covered in this part of the book. Use the research literature and other reading to explore these topics in more depth. Tips to help prepare for topic discussions are available on-line.

13. This chapter has focused upon the amino acids that make up proteins, which are joined together through translation. Discuss the journey of an individual amino acid through the process of translation, starting from its association with tRNA, including the enzymes and protein complexes involved in joining this amino acid to the growing peptide chain, ending with its ultimate release from the ribosome.

14. In Chapter 3, mutations were discussed, including single-nucleotide polymorphisms where a single DNA base is changed to another base. Explore the ways in which the codons that encode the various amino acids could be changed to encode a different amino acid by changing just one base. How does this influence the characteristics of the amino acid that would be added into the protein by the mutated codon?

15. Make a table to classify the various amino acids by their common characteristics and explore how this might impact the role of these amino acids within proteins. If a mutation resulted in a change of one amino acid for another with the same characteristic, would the protein still function? Can you identify examples of this from research?

Further reading

Almeida P. *Proteins: Concepts in Biochemistry*. 1st ed. Garland Science, CRC Press, Taylor & Francis Group, 2016. ISBN 9780815345022.

Branden CI, Tooze J. *Introduction to Protein Structure*. 2nd ed. Garland Science, CRC Press, Taylor & Francis Group, 1999. ISBN 9780815323051.

Williamson M. *How Proteins Work*. 1st ed. Garland Science, CRC Press, Taylor & Francis Group, 2012. ISBN 9780815344469.

Protein Folding and Structure

Proteins are assembled by the ribosomes as strings of amino acids, joined by peptide bonds. To become functional proteins, whether structural proteins or enzymes, the linear string of amino acids must fold into a three-dimensional protein structure. This structure may then undergo further modification, through the addition of non-amino acid components of the protein. For some proteins, conformational changes in the protein can be associated with the protein's function and location within the cell. In this chapter, the various characteristics of protein folding, structure, and modification are explored.

Primary amino acid structure is the linear sequence of amino acids joined by peptide bonds

As in the previous chapter, the structure of a protein as it comes out of the ribosome during translation is the **primary amino acid structure**. It has the amine end of the first amino acid as the N-terminus. This $-NH_2$ is usually a $-NH_3^+$ due to the zwitterionic characteristic of peptide chains. At the other end of the peptide chain is the carboxyl end of the last amino acid at the C-terminus. This is a $-COOH$ end that is generally $-COO^-$, as the zwitterionic form.

While the amino acid chain is how the protein is first formed, this is not how it remains within the cell. It is able to fold into a structure by virtue of the amino acids, their side chains, and features of the primary amino acid sequence. With structure then comes function.

From a genomics perspective it is relatively easy to predict the primary amino acid structure. This is based on the codons that are encoded in the DNA and transcribed into mRNA to be translated at the ribosome. However, what happens next to the primary amino acid structure is less easy to predict and relies upon an array of factors, not only in the amino acid chain itself, but also in the environment of the cell.

Secondary amino acid structure is a folding of the primary sequence of amino acids

An amino acid chain is able to fold into **secondary amino acid structures**. These are simple structures that are stabilized by the formation of hydrogen bonds between amino acids within the peptide chain. There are two basic secondary amino acid structures, the α **helix** and the β **sheet**.

An α helix structure forms from a linear peptide chain and is stabilized due to strong hydrogen bonds between the –CO of internal carboxyl groups and the –NH of internal amine groups along the chain (**Figure 8.1**). The hydrogen bond forms between an amino acid and then the amino acid that is four along in the peptide chain. These hydrogen bonds stabilize the peptide chain into a helix structure. The side chains of the amino acids involved in the α helix structure are pointing out from the helix. The secondary structure of a protein contributes to its function and cellular location. For example, an α helix of 20 amino acids can span a bacterial membrane, making it an integral membrane protein.

Figure 8.1 **The α helix secondary amino acid structure is formed when a linear chain of amino acids forms into a spiral helix that is stabilized by hydrogen bonding.** Amino acids that are internal to the peptide chain hydrogen bond to other amino acids that are four amino acids farther along in the chain. The –CO carboxyl group of one amino acid will hydrogen bond to the –NH amine group of another, giving stability to the helix structure.

A β sheet structure includes hydrogen bonds to stabilize the structure as well. In the case of β sheets, the hydrogen bonds are between neighboring parts of the peptide chain. The structure of β sheets comes in two forms. An antiparallel β sheet has neighboring hydrogen-bonded peptides in opposite directions (**Figure 8.2**). One strand is orientated with its carboxyl group to amine group and the other goes from its amine group to carboxyl group. In the antiparallel β sheet, the hydrogen bonding is symmetrical with one to one hydrogen bonding. A parallel β sheet has peptide chains that are orientated in the same direction, with the hydrogen bonding being non-symmetrical with one amino acid hydrogen bonding to two other amino acids (see Figure 8.2). As for an α helix, the β sheet secondary structure can also contribute to the function and location of the protein, wherein β sheets can also span the bacterial membrane.

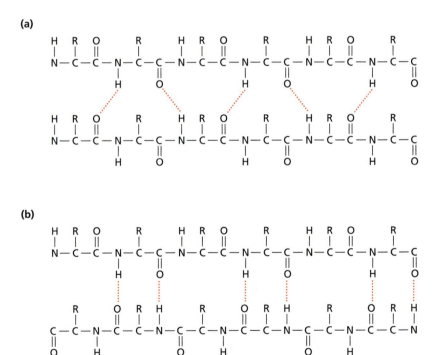

Figure 8.2 **The β sheet secondary amino acid structure is formed when a linear chain of amino acids forms into an aligned sheet of amino acids that is stabilized by hydrogen bonding.** Amino acids that are internal to the peptide chain hydrogen bond to other amino acids that are part of another part of the peptide chain. The –CO carboxyl group of one amino acid will hydrogen bond to the –NH amine group of another, giving stability to the helix structure. (a) In a parallel β sheet, both parts of the chain run in the same orientation. (b) The parts of the β sheet can be antiparallel, with the two parts of the peptide chain running amine group to carboxyl group and the other carboxyl group to amine group.

Bulky side chains tend to interfere with protein secondary structures, so generally α helix and β sheet structures include glycines and alanines, due to the simplicity of their side chain structures. Proline disrupts some secondary structures by virtue of its side chain structure that loops back upon itself. Therefore, proline is often at the end of a helix, bend, turn, or loop, rather than within these structures.

Tertiary amino acid structures form when secondary structures come together

The most complex structure that can form from a string of amino acids is a **tertiary amino acid structure**. Tertiary structures are formed when secondary structures come together with the assistance of several different forces. Hydrogen bonds contribute not only to secondary structures, but also tertiary structures. For example, β sheets can stack with one another and these stacks of β sheets can form a β **barrel**. A β barrel protein structure forms a channel through the center of the β barrel that is frequently seen in membrane proteins, particularly outer membrane proteins in Gram-negative bacteria.

In addition to further interactions through hydrogen bonding, beyond that seen in secondary structures, the side chains of amino acids can interact and form tertiary structures. The forces involved in side chain interactions include ionic interactions and hydrophobic interactions that may result in hydrophobic amino acids being interior to the structure. Tertiary protein structures are also brought about through the disulfide bonds that can form between cysteines, creating bridges within tertiary amino acid structures (**Figure 8.3**). **Oxidative folding** is the process of formation of disulfide bonds between cysteines.

Quaternary amino acid structures form when tertiary structures come together

A single strand of amino acids can form tertiary structures. When two or more tertiary structures come together they form a **quaternary amino acid structure**. Quaternary structures are influenced by hydrogen bonding, hydrophobic amino acids, ionic interactions, and disulfide bonds between two or more tertiary structures. Therefore, the same forces that contribute to tertiary structures are involved in quaternary structures, the difference being that quaternary structures involve more than one protein. Peptides, as folded proteins, come together as dimers, trimers, or more with the resulting structures stabilizing each other further.

There are several examples of proteins that complex with other proteins to form quaternary structures. In each case it may be that these are two copies of the same

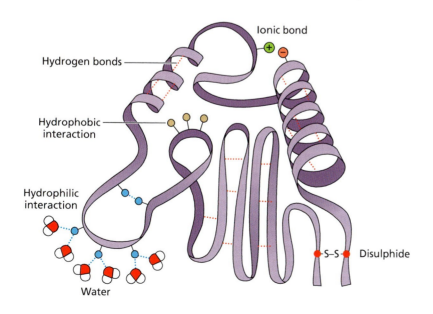

Figure 8.3 Several forces contribute to tertiary amino acid structures. Hydrogen bonding can bring a peptide chain together into secondary structures and can also contribute to tertiary structures. The side chains of the amino acids involved can cause ionic and hydrophobic interactions to occur, contributing to tertiary structure. The presence of cysteines in the protein means there is the potential for disulfide bonds to form within the structure.

Ionic bond

Hydrogen bonds

Hydrophobic interaction

Hydrophilic interaction

Water

S–S   Disulphide

Cytoplasmic membrane   SEC translocase   Periplasm

Cytoplasm

SRP

SecB

Ribosome

Figure 8.18 There are two mechanisms for recognition of the protein to be transported by the Sec translocase. This system transports unfolded proteins that carry a specific amino acid sequence (orange section of peptide) across the cytoplasmic membrane. The signal sequence can be recognized via two mechanisms. In the first, the peptide chain is recognized co-translationally, with the signal recognition particle (SRP) binding to the signal sequence in the peptide chain. Alternatively, a complete yet unfolded peptide chain can be bound by SecB. Both mechanisms then target the peptide for transfer across the cytoplasmic membrane by the Sec translocase.

TAT pathway   Periplasm

Cytoplasmic membrane

Cytoplasm

Folded protein

Figure 8.19 The Tat pathway transports folded proteins across the cytoplasmic membrane. Proteins that are folded in the cytoplasm can be transported across the cytoplasmic membrane by the Tat pathway, also called the twin-Arg pathway. A portion of the protein, the signal sequence (orange section), is recognized by the Tat pathway and targets the protein for transport.

Unfolded proteins can be targeted for transport across the membrane by Sec translocase in two ways, both relying on a **signal sequence** of specific amino acids within the peptide chain. The growing peptide chain can be targeted for transport co-translationally through recognition by the **signal recognition particle (SRP)** as it emerges from the ribosome. Alternatively, after the completion of translation, the signal peptide on the unfolded peptide chain can be recognized by SecB. The unfolded proteins are then transported by the Sec translocase across the membrane and then fold after translocation (**Figure 8.18**).

Proteins can also be translocated across the cytoplasmic membrane in their native folded state through the **Tat pathway** (**Figure 8.19**), also known as a **twin-Arg pathway**. Folded proteins destined to be outside of the cell are translocated across the cytoplasmic membrane of Gram-positive bacteria by these pathways. Likewise, folded proteins destined for the periplasmic space in Gram-negative bacteria can be transported across the inner membrane by the Tat pathway.

Secreted proteins carry a signal to aid in their transport out of the cell

In bacteria such as *E. coli*, nearly half of the proteins are embedded in or cross the bacterial membrane(s). Whether they are transported via the Sec pathway, the Tat pathway, or another mechanism discussed in Chapter 9, most secreted proteins carry a specific amino acid sequence that is used to target the protein for transport. For example, there is a signal sequence of 18–30 amino acids present within the sequence of some proteins that are to be secreted out of the bacterial cell. Generally, secreted proteins have a positively charged N-terminus, usually from lysine or arginine amino acids. These secreted proteins also tend to have a nonpolar hydrophobic core and a polar cleavage region containing a cleavage site. The cleavage site, which breaks the amino acid chain following secretion, is at a cysteine (**Figure 8.20**). The sequence of the features of these proteins is not very well conserved, but the general structure is conserved, with some exceptions.

There are some secreted proteins with no charge or that are negative at the N-terminus and yet they are still translocated. There are also secreted proteins that are not cleaved, which results in them being anchored to the membrane by the transmembrane signal sequence. Therefore, while some secreted proteins can be predicted by sequence data and predicted amino acid features of the protein, others can only be determined experimentally.

Several types of secretion systems have been identified, which also contribute to the transport of proteins, and other components, across the bacterial membrane(s). Some work with the Sec and Tat pathways and some work on their own. These secretion systems are multi-protein complexes that work in a coordinated manner to secrete components out of the bacterial cell. Some activate secretion in response to contact with host cells and other stimuli. The various types of secretion systems will be explored in greater detail in Chapter 9.

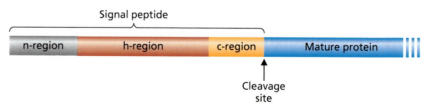

Signal peptide

| n-region | h-region | c-region | Mature protein |

Cleavage site

Figure 8.20 The features of the signal peptide region of a secreted protein. Although the amino acid sequence may vary, the general features of the N-terminus of proteins that are secreted are conserved. First there is an n-region that contains positively charged residues, followed by a hydrophobic h-region that can form an alpha helix, and lastly a c-region. The c-region is the location of the cleavage site, where the signal peptide is cleaved from the mature protein during transport across the membrane.

Key points

- The primary amino acid structure is the chain of amino acids as they come out of the ribosome during translation, but this is not their final form in the bacterial cell.

- Peptide chains can form α helices and β sheets as secondary amino acid structures, both of which rely on hydrogen bonding to form the structure.

- Three-dimensional tertiary structures of peptide chains vary, but can be the result of β sheets stacking into β barrels, or from amino acid side chains interacting with one another.

- Two or more tertiary protein structures can come together in a quaternary structure, such as seen in RNA polymerase with its subunits.

- Protein folding can be aided by chaperone proteins, which aid in the proper folding of proteins and the resolution of misfolding.

- Many post-translational modifications can add molecules to the amino acids that make up a bacterial protein.

- Modifications to proteins such as phosphorylation or lysine acetylation are a reversible means to activate and inactivate protein functions.

- The first amino acid in a peptide chain is *N*-formylmethionine; post-translationally the formyl group can be removed or the whole *N*-formylmethionine can be removed.

- Proteins can be transported across the cytoplasmic membrane unfolded by the Sec pathway or folded by the Tat pathway.

Terms, questions, and discussions

Key terms

Define the following terms introduced in this chapter. Check your answers using the definitions in the Glossary. These terms are also available as Flashcards on-line.

α helix	Hydroxylation	Quaternary amino acid structure
β barrel	Lipidation	*S*-nitrosylated
β sheet	Lipoprotein	Sec pathway
Acetyl CoA	Lysine acetyltransferase	Sec translocase
Acetylation	Lysine deacetylase	Secondary amino acid structure
Acylation	Methionine aminopeptidase	Signal recognition particle
Chaperones	Methylation	Signal sequence
Deformylase	N-terminal acetyltransferase	Succinylation
Glycoprotein	Phosphorylation	Tat pathway
Glycosylation	Primary amino acid structure	Tertiary amino acid structure

Self-study questions

Answer each question using 50–100 words or a table or labeled diagram. Advice on where to find answers to these questions is available on-line.

1. When an amino acid emerges from the ribosome during translation, which end is first, the amine group or the carboxyl group, and how does this reflect the terminus name of the peptide chain?

2. What are the two basic secondary structures that can be formed by amino acids and what stabilizes the formation of each of these structures?

3. Which tertiary amino acid structure forms from the stacking of secondary amino acid structures? How does the nature of this structure contribute to its function when part of a membrane protein?

4. What are the key differences between tertiary amino acid structures and quaternary amino acid structures?

5. Tertiary structures can include bridges within them formed by disulfide bonds. Which amino acids are involved in these bonds and what is the name of the process of formation of these bonds? Which proteins assist in the correct formation of disulfide bonds?

6. Which processes within a bacterial cell are influenced by phosphorylation of proteins? Which amino acids within proteins can be modified to add a phosphate?

7. What does the addition of a lipid to a protein do to the nature of the protein? How is it possible to recognize a protein that can be lipidated?

8. What addition do glycoproteins have added to the amino acid(s) and how does this influence the nature of the protein? What biological processes are glycoproteins involved in for bacteria?

9. Draw the structures involved when acetyl-CoA acts on a peptide chain, resulting in N-terminal protein acetylation.

10. Draw a lysine residue and then the two potential acylations of lysine.

11. Lysine is also modified by methylation. Add to the drawing above a lysine that has been methylated. What cellular processes in bacteria include methylated proteins?

12. What modifications can occur to the first amino acid that is added to a peptide chain?

Discussion topics

These topics are presented for discussion in study groups, as part of class discussions, or on your own. These questions go beyond what is directly covered in this part of the book. Use the research literature and other reading to explore these topics in more depth. Tips to help prepare for topic discussions are available on-line.

13. In this chapter the structures of proteins have been explored, including internal bonds, modifications, and interactions between proteins and subunits. Using protein structural databases, investigate the structure of a protein of interest, identifying the presence of any α helices, β sheets, β barrels, hydrogen bonds, disulfide bonds, and post-translational modifications. Is there evidence in the research literature that chaperones are involved in the folding of this protein or other proteins involved in its modification?

14. Post-translational modifications, such as phosphorylation, can play a vital role in gene regulation, as seen with two-component regulatory systems, and influence such traits as virulence. Discuss the role of any post-translational modification of a specific protein or set of proteins that directly influences the virulence of a bacterial pathogen.

15. Protein transport and secretion is essential for the biological functions of the bacterial cell. Using a species of interest as an example, investigate those proteins that are transported via the Sec pathway and those that are transported via the Tat pathway. Compare and contrast the types of proteins transported by each pathway, the features of each protein, their final destinations, and the presence of any signal sequences.

Further reading

Protein folding

De Geyter J, Tsirigotaki A, Orfanoudaki G, Zorzini V, Economou A, Karamanou S. Protein folding in the cell envelope of *Escherichia coli*. *Nat Microbiol*. 2016 July 26; 1(8): 16107.

Post-translational modification

Ambler RP, Rees MW. Epsilon-*N*-methyl-lysine in bacterial flagellar protein. *Nature*. 1959; 184: 56–7.

Bastos PA, da Costa JP, Vitorino R. A glimpse into the modulation of post-translational modifications of human-colonizing bacteria. *J Proteomics*. 2017 January 30; 152: 254–75.

Buddelmeijer N. The molecular mechanism of bacterial lipoprotein modification—How, when and why? *FEMS Microbiol Rev*. 2015 Mar; 39(2): 246–61.

Eichler J, Koomey M. Sweet new roles for protein glycosylation in prokaryotes. *Trends Microbiol*. 2017 Aug;25(8):662–672. doi: 10.1016/j.tim.2017.03.001.

Lanouette S, Mongeon V, Figeys D, Couture J-F. The functional diversity of protein lysine methylation. *Mol Syst Biol*. 2014 Apr 4; 10(4): 724.

Ouidir T, Kentache T, Hardouin J. Protein lysine acetylation in bacteria: Current state of the art. *Proteomics*. 2016 Jan; 16(2): 301–9.

Ren J, Sang Y, Lu J, Yao YF. Protein acetylation and its role in bacterial virulence. *Trends Microbiol*. 2017 Sep;25(9):768–779.

Figure 9.1 Lipopolysaccharide (LPS) makes up the outer leaflet of the outer membrane of Gram-negative bacteria. It is one of several non-protein components of the Gram-negative bacterial cell, including also the phospholipids of the inner membrane and inner leaflet of the outer membrane and the peptidoglycan of the cell wall, present in the periplasm.

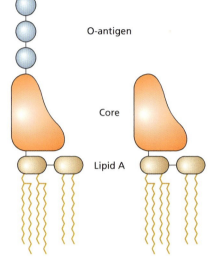

Figure 9.2 The three-part structure of LPS. Lipopolysaccharide has three components: lipid A (beige); core (orange); and O-antigen (blue). The lipid A portion anchors the large LPS molecule in the outer surface of the Gram-negative outer membrane. The core and O-antigen portions of the molecule extend out from the bacterial cell into the surrounding environment. Most Gram-negative species have the longer LPS (left), while a few have shorter LOS on their surface (right), which does not have O-antigen.

Figure 9.3 Lipid A genes. Some of the genes responsible for the production of the lipid A portion of LPS are present together in clusters, such as *lpxD*, *lpxA*, and *lpxB* shown here, facilitating the coordinated expression of the various enzymes needed for biosynthesis of lipid A.

The **lipid A** portion of LPS is highly conserved between the various Gram-negative bacterial species. As such, the genes encoding the production of lipid A are likewise highly conserved. The genes involved in the synthesis of lipid A are *lpxA*, *lpxC*, *lpxD*, *lpxH*, *lpxB*, *lpxK*, *kdtA*, *lpxL*, and *lpxM* (**Figure 9.3**). Some of these genes are present together in clusters, which aids in the coordination of expression of the gene products and therefore the coordinated production of lipid A. The *lpxD-fabZ-lpxA-lpxB* cluster includes the *lpx* genes involved in lipid A biosynthesis and also the gene *fabZ*, which is involved in production of phospholipids. The presence of genes for these two different biosynthetic processes together in one gene cluster may perhaps balance the production of these two essential non-protein membrane components.

The **LPS core** is an oligosaccharide made up of 10–15 sugars that is connected directly to the lipid A part of the lipopolysaccharide. The core itself is made up of the inner core, the outer core, and the kdo portion. These parts of the core are made using the enzymes encoded in three separate operons. The *waaD-waaF-waaC-waaL* gene cluster (**Figure 9.4**) is involved in the biosynthesis of the inner core oligosaccharide. The *waaQ* operon (see Figure 9.4) is adjacent to the other cluster and has seven to 10 genes, the products of which are responsible for production of the outer core. Finally, the *kdtA* operon gene products add the kdo residue to the LPS core.

The portion of the LPS that is farthest from the bacterial cell is the **O-antigen**, which extends out into the environment. The O-antigen is made up of a repetitive glycan polymer of one to eight sugars that is attached to the LPS core. The structure, length, and characteristics of the O-antigen varies, dependent upon the bacterial species and strain. In *Escherichia coli*, for example, there are over 160 known different O-antigens in the various strains of this species. The phenotypic characteristics that the LPS confers upon a bacterial colony growing on solid agar media has resulted in full-length LPS possessing strains being called "**rough**," while LPS with shorter O-antigen is referred to as "**smooth**." The length and features of the O-antigen of LPS can have an impact upon the bacteria's resistance to serum killing and antibiotics. Because it is a differentiating characteristic between strains, the O-antigen is the basis of serotyping in some species, enabling different strains with different LPS structures in the O-antigen to be identified, including strains of *E. coli*, *Yersinia*, and *Vibrio*. The variation in O-antigen means there is a great deal of variation in the genes that will be involved in its biosynthesis, however, most are part of the *rfb* gene cluster (**Figure 9.5**). Some species, such as *Neisseria* and *Haemophilus*, have no O-antigen in their LPS, where it is often referred to as LOS, reflecting its shorter structure with no O-antigen (see Figure 9.2).

Figure 9.4 The operons for LPS core production. The LPS core *waaD* and *waaQ* operons are adjacent to one another yet transcribed in opposite orientations (arrows). The *waaD-waaF-waaC-waaL* genes of the *waaD* operon (light orange) encode the proteins that make the inner core. The convergently transcribed *waaQ* operon (dark orange) is responsible for the outer core and includes up to 10 genes. The kdo residues of the LPS core are added by the products of the *kdtA* operon (not shown).

| galF | rmlB | rmlD | rmlA | rmlC | wzx | mnaA | wekM | wzy | wekN | wekO | | gnd |

| galF | rmlB | rmlD | rmlA | fdtABC | wzx | wekP | rmlC | wzy | wekQ | wekR | yhhI | wekS | gnd |

| galF | rmlB | rmlD | rmlA | rmlC | wzx | | | wzy | wekT | wekU | wekV | wekN | gnd |

Figure 9.5 Genetics of the LPS O-antigen. The genes responsible for production of the O-antigen of LPS in *E. coli* are present between *galF* and *gnd*. The genes within this region vary between strains of *E. coli*, as shown in these three example regions. In the top region are *mnaA*, *wekM*, *wekN*, and *wekO*, which are not in the other two strains. In the middle, genes *fdtABC* replace *rmlC* between *rmlA* and *wzx* in the order of the region. The *rmlC* gene is present in this strain, however it is later, after the unique gene *wekP*. In the bottom region there are no genes between *wzx* and *wzy*, however, like the other strains there are different genes between *wzy* and *gnd*.

The LPS has to be assembled and translocated to the outer surface of the cell

Components of the LPS are made within the bacterial cell, yet their final place is on the outer surface of the bacteria. LPS synthesis starts in the cytoplasm with lipid A, which must therefore be translocated across the membranes. Lipid A is hydrophobic, therefore, it begins embedded in the inner membrane. It is passed cross to the periplasmic side of the inner membrane. It is here that the core and O-antigen are added. Subunits of O-antigen are also made in the cytoplasm where they are polymerized by the Wzy protein. The chain length of O-antigen repeats is determined by Wzz, an inner membrane protein. The O-antigen subunits are then translocated across the inner membrane by the Wzx membrane protein. Repeat chains of O-antigens are ligated to the LPS core and lipid A on the outer surface of the inner membrane by the protein WaaL, which is also present in the inner membrane. Then the complete LPS is translocated from the periplasmic aspect of the inner membrane to the outer leaflet of the outer membrane by the proteins LptA, LptB, LptC, LptD, LptE, LptF, and LptG. As with the biosynthesis of LPS, the genes responsible for translocation of LPS tend to be present on the genome in gene clusters (**Figure 9.6**).

Peptidoglycan is built by proteins encoded in a cluster of genes

Similarly to LPS, the components of the peptidoglycan cell wall are made in the cytoplasm. The products of the genes *murABCDEF*, *mraY*, and *murG* act, in this order, to generate peptidoglycan precursors in the cytoplasm (**Figure 9.7**). These precursors are then translocated to the periplasm where the repeating units of **N-acetyl glucosamine** (**GlcNAc** or **NAG**) and **N-acetyl muramic acid** (**MurNAc** or **NAM**) are assembled. The MurNAc components include peptide chains that add to the structural integrity of the peptidoglycan by cross-linking the layers of the cell wall.

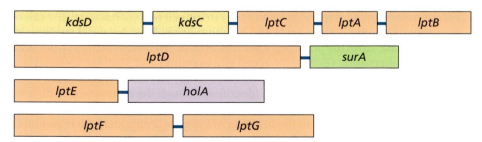

Figure 9.6 Genes that are responsible for translocation of LPS. Several of the genes encoding proteins involved in the translocation of LPS components from the cytoplasm to the periplasmic aspect of the inner membrane are present together in the chromosome. These include the *kdsD-kdsC-lptC-lptA-lptB* genes, the *lptD-surA* genes, *lptE-holA* genes, and the *lptF-lptG* genes. Together the products of these genes come together to form the LptABCDEFG multi-protein system responsible for the translocation of LPS from the inner membrane to the outer leaflet of the outer membrane.

Figure 9.10 Different thicknesses of peptidoglycan in Gram-positive (left) and Gram-negative (right) bacteria. The Gram-positive peptidoglycan layer, which is on the outside of the cell, is very thick and can make up to 90% of the weight of the cell. There is a thin layer of peptidoglycan in Gram-negative bacteria, where the cell wall makes up about 20% of the weight of the cell.

Bacterial membrane phospholipids are made by proteins

Cellular membranes are made up, for the most part, of phospholipids, with their hydrophilic polar heads and hydrophobic non-polar tails (**Figure 9.11**). In order for the bacterial cell to grow and divide, more phospholipids must be generated. To adapt to changing environmental conditions, bacteria respond by changing their membrane phospholipids. As discussed in the previous chapter, such changes to the membrane can result in changes in exposure and activity of membrane proteins. Changes in the membrane can protect the bacterial cell from stress and may change during biofilm formation. Genomics has revealed that many of the genes involved in making non-protein cellular components of the bacterial cell, such as the biosynthesis of LPS and the generation of peptidoglycan synthesis, are conserved between species. For these, *E. coli* is seen as the representative species for such genetic investigations. Other cellular processes and the genes involved, such as genes responsible for the production and modification of phospholipids, differ between species. Regardless of whether the genes are conserved or unique to a specific bacterial species, the genes encode proteins, which are enzymes that in turn produce the phospholipids that make up the bacterial membrane. The genes themselves do not directly encode the phospholipids, yet without the genes there would be no phospholipids.

The lipid bilayer of a bacterial cellular membrane is 6–8 nm thick and is made of two layers of phospholipid molecules (**Figure 9.12**). The bacterial membrane in Gram-positive bacteria and the inner membrane in Gram-negative bacteria are made of mostly phospholipids. Within this membrane are α-helical proteins as well, however greater than 95% of the membrane lipids are the phospholipids making up the lipid bilayer.

Gram-negative bacteria have a second cellular membrane, which also contains phospholipids. The outer membrane has both phospholipids and LPS making the membrane bilayer, rather than a bilayer of phospholipids as in the inner membrane (**Figure 9.13**). The proteins in the Gram-negative outer membrane tend to have β-barrel structures, unlike the α-helical proteins of the inner membrane. Lipoproteins are anchored by N-terminal acyl modifications and can be found in both membranes.

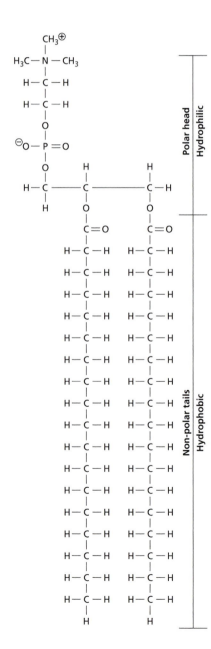

Figure 9.11 Structure of a phospholipid. The phospholipid has two parts: a hydrophilic polar head; and a hydrophobic non-polar tail. The head includes a phosphate group. The bonds between the carbons in the tails can change from single bonds as shown here to double bonds (not shown), causing bends in the tail chains.

Figure 9.12 Structure of a phospholipid bilayer. The bacterial cytoplasmic membrane is made almost entirely of phospholipids. The hydrophilic heads form the surface of the membrane, while the hydrophobic tails form the center of the membrane. The membrane can include embedded α-helical proteins.

Figure 9.13 The outer membrane of the Gram-negative bacterial cell. The outer membrane of Gram-negative bacteria is made up of both phospholipids and LPS. Most of the outer leaflet of the outer membrane is LPS. Embedded proteins tend to be β-barrel proteins.

Extracellular polysaccharides make up the bacterial capsule

Some bacterial cells are covered in a **capsule**. The presence of a bacterial capsule is often important for virulence and survival within the host, enabling the bacteria to avoid being killed by the immune response. Capsules can also protect the bacterial cell from drying out, therefore, being an important feature of species like *Staphylococcus epidermidis* that live on the surface of the skin. Previously, the experiments conducted that transferred virulence to non-virulent bacteria were discussed, having been used to demonstrate that DNA is the genetic material. The feature that differentiated these virulent and non-virulent bacteria was the capsule, present in the virulent bacteria.

Bacterial capsules are made of extracellular polysaccharides that are anchored to the cell membrane through the lipid component of glycolipids. The capsule is present outside of the outer surface of the Gram-negative outer membrane, outside of the LPS, or beyond the thick peptidoglycan of the Gram-positive bacteria (**Figure 9.14**).

As with many of the other non-protein features of the bacterial cell, the genes responsible for the biosynthesis of the capsule tend to be present in clusters, close to each other on the genome. Changes in the repertoire of these genes, such that different genes are found in otherwise similar strains, influences the formation and structure of the capsule and therefore influences the strain-to-strain variation in the capsule. These differences in capsule types are often reflected in differences in serogroups, which are used to differentiate bacterial strains phenotypically. It has long been known that the DNA encoding capsule genes is able to transform bacteria of the same species that either do not have capsule genes, resulting in capsulate strains, or transforming those that do have capsule genes to change the characteristics of the capsule produced. Horizontal gene transfer has been recognized with regards to bacterial capsule genes for quite some time, due to the easily identifiable change in phenotype.

Figure 9.14 The bacterial capsule. Some bacterial cells are surrounded by a layer of extracellular polysaccharides that form the bacterial capsule. The capsule is visible here as the dark blue layer surrounding the bacterial cells. The capsule can help the bacteria avoid the immune system and resist drying. (Courtesy of Dr Kari Lounatmaa/Science Photo Library.)

Proteins make up bacterial cell structures

Bacterial proteins are more than just the enzymes involved in the biosynthesis of non-protein cellular structures and cellular metabolism. Proteins also themselves form bacterial cellular structures, some of which can be visible microscopically, such as **pili**, **fimbriae**, and **flagella** (**Figure 9.15**). Pili are long thin structures that protrude from the surface of the bacterial cell, made up of protein subunits. These subunits are assembled together into the long pilus strand by pilus machinery, which is also made of protein. Pili can then be modified by enzymes, often changing the **antigenicity** of the pili, its ability to be recognized by the immune system.

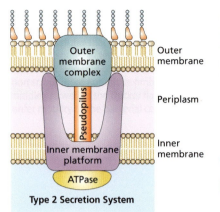

Figure 9.17 Type 2 Secretion System.
A Type 2 Secretion System has four parts that cross the inner membrane, periplasmic space, and outer membrane. Embedded in the inner membrane is the inner membrane platform (purple) with the secretion ATPase (yellow), which powers the secretion through ATP. The outer membrane complex (green) forms a channel in the outer membrane through which the protein is secreted with the help of the pseudopilus (orange).

transport from ATP within the cell. The T2SS pseudopilus is genetically related to pili, the long thin protein structures present in some species that extend from the surface of the bacterial cell. It is believed that the pseudopilus retracts and then pushes the folded protein through the T2SS outer membrane complex (**Figure 9.17**), in a manner similar to the retraction of pili in twitching motility.

Type 2 Secretion Systems tend to secrete proteases, lipases, and phosphatases. *Vibrio cholerae* uses a T2SS to secrete cholera toxin, characteristic of this species. Another pathogen, *Pseudomonas aeruginosa*, secretes exotoxin A via a T2SS.

The Type 3 Secretion System can inject proteins like a syringe

The **Type 3 Secretion System** (**T3SS**) is particularly prevalent in Gram-negative pathogens and symbionts. These bacteria have a needle and syringe structure that is similar to the bacterial flagella. The Type 3 Secretion System is able to secrete proteins across the inner membrane and the outer membrane. Some T3SSs transport proteins across both of the Gram-negative bacterial membranes and also across the eukaryotic membrane. In this way, T3SSs are able to inject proteins directly into host cells, delivering bacterial proteins directly to their targets. Proteins secreted by T3SSs are called **effectors** and have a secretion signal for Type 3 Secretion within their N-terminus.

The process of secretion via a Type 3 Secretion System is ATP dependent, as with the Type 1 and Type 2 Secretion Systems. T3SSs transport unfolded proteins, which pass as peptide chains through the needle and syringe structure of the transport system. There is a core of nine proteins that are highly conserved in all T3SSs; eight of these are evolutionarily similar to proteins in flagella. There are an additional 10–12 other proteins that contribute to the T3SS.

Operons with T3SS genes are typically found in pathogenicity islands, horizontally transferred DNA of the genome or plasmid. There are three parts to T3SS: the **base complex**; the **needle**; and the **translocon** (**Figure 9.18**). The base complex, also called the **basal body**, has both cytoplasmic and outer membrane parts. The structure of the base complex includes several rings and a central rod, made up of about 15 proteins. The Type 3 Secretion System needle comes out of the basal body. The needle has a hollow core that is big enough for the passage of an unfolded protein through it. At the end of the needle is the translocon, which is able to sense contact with the host cell and regulates secretion of the effectors based on contact with the host cell. In species that transport effectors directly into host cells, the translocon assembles on contact with the host cell and forms a pore in the host cell membrane.

The Type 4 Secretion System includes conjugation systems and DNA uptake systems

The **Type 4 Secretion System** (**T4SS**) includes the conjugation systems for transfer of DNA between bacterial cells, thus there is homology between the T4SS and pili. While the conjugation system is involved in transfer of DNA involving cell-to-cell contact, the T4SSs can transport proteins, protein–protein complexes, and DNA–protein complexes either between cells or between the cell and the environment. T4SSs can secrete these substrates into other cells, be they bacterial or eukaryotic cells, and can secrete them out of the cell or be involved in uptake. Transport via T4SSs is through the inner and outer membrane and can also go through another membrane, like the T3SSs (**Figure 9.19**).

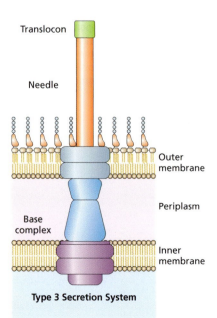

Figure 9.18 Type 3 Secretion System. A Type 3 Secretion System has three parts that cross the inner membrane, periplasmic space, outer membrane, and extend out of the cell. Embedded in the inner membrane, crossing the periplasm, and embedded in the outer membrane is the base complex, made up of several proteins that have homology to flagellar proteins. Extending out of the cell is the Type 3 Secretion System needle, which has at its end the translocon.

Figure 9.19 Type 4 Secretion System. A Type 4 Secretion System crosses the inner membrane, periplasmic space, and outer membrane. Several proteins make up the structure that spans and connects both membranes, providing a channel through which secretion substrates can pass. The Type 4 Secretion System proteins have homology to proteins of the pili and pilus assembly machinery, with those involved in conjugation making up the conjugation pilus.

Type 5 Secretion Systems are proteins that secrete themselves

In a **Type 5 Secretion System** (**T5SS**), there is no separate membrane channel or secretion apparatus. A T5SS protein secretes itself out of the cell. T5SS proteins have their own β-barrel domain that inserts into the outer membrane to form a pore through which the protein is secreted. A Type 5 Secretion System is only able to provide transport for a protein through the outer membrane. The T5SS proteins are brought into the periplasm as unfolded proteins secreted by the Sec pathway and, therefore, they have N-terminal Sec signal sequences.

There are three types of Type 5 Secretion Systems. The first involves proteins that are called **autotransporters** because they can transport themselves. The autotransporters are the most simplistic of the T5SSs. The transporter domain is in the C-terminal end of the protein and forms the outer membrane channel, made up of a 12-stranded β-barrel. An example of a T5SS autotransporter is the immunoglobulin A protease from *Neisseria gonorrhoeae*, which cleaves human IgA antibodies. The second type of T5SSs is the **two partner secretion**, where a pair of proteins are involved. In two-partner secretion, one protein forms the β-barrel outer membrane channel and the other is the secreted protein. Lastly, **chaperone-usher secretion** involves three proteins. The usher component of chaperone-usher secretion forms the β-barrel in the outer membrane. The chaperone in the periplasm folds the Sec pathway transported secretion protein and delivers it to the usher. The third component of the chaperone-usher secretion T5SS is the secreted protein itself (**Figure 9.20**).

Type 6 Secretion Systems transport proteins into other cells, including other bacteria

The most recent Gram-negative secretion system described is the **Type 6 Secretion System** (**T6SS**). Proteins can be transported from the bacterial cell into other cells, including other bacteria, fungi, and eukaryotic cells via the T6SS. The proteins that make up the T6SS are encoded within a gene cluster of up to 21 genes. The proteins of the T6SS are structurally homologous to bacteriophage tails (**Figure 9.21**). The

Type 4 Secretion System

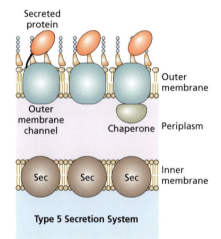

Type 5 Secretion System

Figure 9.20 Type 5 Secretion System. A Type 5 Secretion System describes a protein that transports itself out of the periplasm. There are three types of Type 5 Secretion Systems, each relying on the Sec pathway (Sec) in the inner membrane to deliver the secreted protein to the periplasm. Autotransporters (left), form an outer membrane channel (green) in the outer membrane, through which the secreted portion (orange) of the autotransporter protein passes. In two partner secretion (middle), two different proteins are the outer membrane channel and the secreted protein. Chaperone-ushered secretion (right), has a third component (green) that chaperones the folding of the secreted protein in the periplasm.

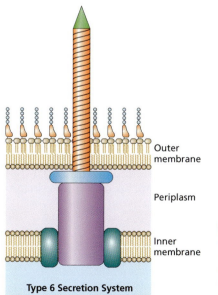

Type 6 Secretion System

Figure 9.21 Type 6 Secretion System. A Type 6 Secretion System crosses the inner membrane, periplasmic space, and outer membrane. Several proteins that are homologous to bacteriophage make up the structure that spans and connects both membranes, providing a channel through which secretion substrates can pass.

T6SSs contribute to interspecies competition, where one bacteria can inject another with proteins that will kill it.

The T6SS is a contractile nanomachine with a puncturing structure at the tip. The contractile mechanism of the T6SS is similar to the tail and spike of bacteriophage. The tip of the T6SS is made of the VgrG protein, which associates with various specialized effector proteins that are toxic to competitor cells found in the bacterial cell environment. In addition, there are cargo effector proteins that are not at the tip of the T6SS. Effectors include peptidoglycan hydrolases that breakdown bacterial cell walls, phospholipases that degrade phospholipids, DNases, and pore-forming toxins. For each effector protein, the bacterial cell has an immunity protein to protect it from the antimicrobial effects of its own T6SS toxins. For example, the Ssp1 and Ssp2 antibacterial toxins are encoded in the T6SS gene cluster, as are their corresponding immunity proteins Rap1a and Rap2a.

Gram-positive secretion systems can aid protein transport across the thick peptidoglycan layer

Gram-positive bacteria have the Sec and Tat pathway to translocate proteins across the membrane. To achieve more efficient protein secretion, many Gram-positives have two *secA* genes encoding the Sec pathway protein. The second of these genes encodes a protein that participates in the transport of only select proteins. Despite these systems, it is often not enough for a protein to escape the cell. In part this is due to the thick peptidoglycan layer, although proteins can eventually be secreted out by passive diffusion.

Type 7 Secretion Systems (**T7SS**) have been described in Gram-positive bacteria. These systems are able to transport proteins through both the membrane and the peptidoglycan layer, reducing the reliance on passive diffusion to secrete the protein out of the cell (**Figure 9.22**). The T7SSs were first identified in *Mycobacteria tuberculosis* and *Staphylococcus aureus*, where the genes encoding the Type 7 Secretion System are present in clusters.

Efflux pump systems transport harmful substances out of the bacterial cell

Efflux pump systems are multiprotein systems that remove substances from the periplasm and/or cytoplasm that are damaging to the bacterial cell. Efflux pumps can play an important role in antibiotic resistance through actively removing antibiotics from the bacterial cell. The transport involved in efflux pump systems is powered by ATP or by the **proton motive force** (**PMF**). Proton motive force results from cellular metabolism. Protons that are not used in the bacterial cell are exported to the surface and either interact with the LPS or the cell wall. These protons on the surface of the cell form an electrochemical gradient across the membranes that can be harnessed as an energy source for efflux pump transport.

There are five families of efflux pumps. The first are the **RND family** of efflux pumps. RND stands for resistance nodulation division, representing the cellular processes in which the RND proteins were believed to be involved. RND efflux pumps are only present in Gram-negative bacterial species. The classic RND efflux pumps in *E. coli* are AcrAB-TolC and AcrEF-TolC. RND family efflux pump systems are made up of three proteins: AcrB, which is in the inner membrane; AcrA, which is a membrane fusion protein that aids movement through AcrB; and TolC, which is the outer membrane channel that connects with AcrB (**Figure 9.23**). The protein TolC is a general outer membrane pore protein that is involved with other protein complexes, such as the T1SS, as discussed earlier.

Figure 9.22 Type 7 Secretion System. A Type 7 Secretion System crosses the Gram-positive membrane and the thick peptidoglycan layer, helping to efficiently secrete proteins out of the cell.

Figure 9.23 RND Efflux Pump. The RND efflux pump of Gram-negative bacteria crosses the inner membrane, periplasmic space, and outer membrane. It is made up of an inner membrane protein, a membrane fusion protein, and an outer membrane protein. These proteins work together using the proton motive force to remove substances from the cytoplasm and periplasm of the bacterial cell.

The second family of efflux pump systems is the **MFS family**. MFS stands for major facilitator superfamily. This type of efflux pump is present in both Gram-negative and Gram-positive bacterial species. The efflux pump EmrD in *E. coli* is a member of the MFS family. It has 12 transmembrane helices in the inner membrane that form a channel for transport of hydrophobic substrates out of the bacterial cell using the proton motive force (**Figure 9.24**).

The third efflux pump family is the **ABC family**. In this case ABC stands for ATP-binding cassette. In this efflux pump system, there is a protein that binds to and uses ATP as the energy source for transport of substrates out of the cell (**Figure 9.25**). The ABC family efflux pumps are present in both Gram-negative and Gram-positive bacterial species.

The fourth family of efflux pump systems is the **SMR family**, which uses the proton motive force. These are classified as small multidrug resistance, SMR, and are present in both Gram-negative and Gram-positive bacterial species. The efflux pump system EmrE in *E. coli* is part of the SMR family. This efflux pump system includes eight EmrE proteins, each with four transmembrane helices (**Figure 9.26**).

The fifth and final efflux pump system family is the **MATE family**, where MATE stands for multidrug and toxic compound extrusion (**Figure 9.27**). MATE family efflux pumps are found in both Gram-negative and Gram-positive bacterial species and make use of sodium influx to drive efflux of harmful substances.

All of the expressed proteins are the proteome

The bacteria genome is relatively stable. Every protein that can be made is there, encoded in the genes, but the genome does not represent which proteins are expressed and when. The bacterial transcriptome indicates the potential of protein production more than the genome. The mRNA within the bacterial cell gives an indication of protein expression. Yet, only the proteome shows the proteins that are currently present and presumably active in the bacterial cell at the time of assessment.

Proteomics analyzes all of the proteins. Like a transcriptome, the proteome changes over time and due to changes in cellular conditions. When conditions shift, the change in the proteome may be either more or less rapid than the changes in the transcriptome. This is due to the relative half-life and enzymatic degradation of the proteins and the mRNA molecules in the bacterial cell.

Through proteomics, it is possible to identify the protein products of genes annotated in the genomes as hypothetical. Comparative proteomics, where the proteins expressed in one condition are compared to the proteins expressed in another, can determine differences in the patterns of protein expression in various conditions.

Whereas sequence technologies made genomics and transcriptomics possible, **mass spectrometry** made proteomics possible. Previously, the study of proteins had been accomplished as standard using **2D gels**. The resolution of proteins in 2D gels works through the separation of proteins by **isoelectric point** in the first dimension and then the protein **molecular weight** in the second dimension. Bacterial cellular proteins are resolved on 2D gels as spots that can be extracted from the gel and identified by mass spectrometry. Although the combination of 2D gels and mass spectrometry of spots is powerful and has revealed a great deal about the proteome, the resolution of 2D gels makes it difficult to identify low abundance proteins and those with high or low molecular weights. In addition, each individual spot of protein must be selected and extracted from the gel for mass spectrometry analysis, which is a labor-intensive process prone to introduction of contaminants. If there are few changes between two conditions and therefore relatively few spots to be extracted and prepared for mass spectrometry, this can be achieved. However, as the number of changes in protein expression increases, so does the difficulty in using this method for proteomic investigations.

Mass spectrometry technology enables the study of proteomes

Gel-free mass spectrometry technology, aka shotgun proteomics, has expanded the ability to study proteomes in much the same way as next-generation sequencing has

Figure 9.24 MFS efflux pump. The MFS efflux pump of Gram-negative bacteria crosses the inner membrane or the cellular membrane in Gram-positive bacteria.

Figure 9.25 ABC efflux pump. The ABC efflux pump of Gram-negative bacteria crosses the inner membrane or the cellular membrane in Gram-positive bacteria.

Figure 9.26 SMR efflux pump. The SMR efflux pump of Gram-negative bacteria crosses the inner membrane or the cellular membrane in Gram-positive bacteria.

Figure 9.27 MATE efflux pump. The MATE efflux pump of Gram-negative bacteria crosses the inner membrane or the cellular membrane in Gram-positive bacteria.

Such investigations have made it clear that in several species there are multiple overlapping resistance mechanisms that may all be at work within the same strain.

MALDI-TOF-MS is less expensive and more rapid than plate culture-based resistance determination, which takes several additional days and additional labor and human decision making. MALDI-TOF-MS produces results based on expressed proteins unlike molecular methods relying on DNA or RNA, which can only identify the presence of the gene or transcript, not the protein that is ultimately the resistance determinant. The advantage in identifying antibiotic resistance using proteomics is that it is known that the protein is expressed. The bacterial cell both has the gene conferring resistance and it is using it. Unfortunately, proteomics alone, like molecular tests for DNA or RNA, does not give MIC data. The Minimum Inhibitory Concentration of the antibiotic to kill the particular bacterial strain contributes to epidemiology data about changes in resistance levels of bacterial species over time. In addition, this data is important for monitoring of resistance locally and globally and for making recommendations about prescriptions for effective management of disease with antibiotics.

Key points

- Some parts of a bacterial cell are not made of protein, such as phospholipids, lipopolysaccharide, peptidoglycan, and the bacterial capsule.

- The biosynthesis of non-protein components of the bacterial cell relies on proteins.

- Pili, fimbriae, and flagella are bacterial structures made of proteins that extend from the surface of the cell.

- Six types of secretion systems have been identified in Gram-negative bacteria and one type has been identified in Gram-positive bacteria, which facilitate the passage of proteins out of the cell.

- Some secretion systems have evolved from other bacterial systems, such as pili, flagella, and bacteriophage.

- There are five families of efflux pump systems that actively remove substances from bacterial cells, including antibiotics.

- Proteomics is the study of the proteins within a bacterial cell, revealing all of the proteins encoded by the DNA that have been transcribed and translated into protein.

- Improvements in mass spectrometry have led to improvements in the results that can be generated in proteomic investigations.

- Proteomic technology can be used to identify bacteria.

Terms, questions, and discussions

Key terms

Define the following terms introduced in this chapter. Check your answers using the definitions in the Glossary. These terms are also available as Flashcards on-line.

2D gel	LPS	T2SS
Autotransporter	MALDI-TOF-MS	T3SS
Capsule	Mass spectrometry	T4SS
Efflux pump system	Pili	T5SS
Flagella	Proteomics	T6SS
HPLC	Proton motive force	T7SS
LC/MS/MS	T1SS	Twitching motility

Self-study questions

Answer each question using 50–100 words or a table or labeled diagram. Advice on where to find answers to these questions is available on-line.

1. In what ways does the proteome relate to the genome and the transcriptome?

2. What are some of the non-protein components of bacterial cells? How are these biosynthesized within the bacterial cell?

3. What kind of bacterial cells have LPS and what are the three parts of the LPS molecule? Briefly describe the ways LPS can vary between strains of the same species.

4. What are the components that make up peptidoglycan for most bacterial species? Which of these is specific to bacteria, not being found elsewhere in nature?

5. In what circumstances might temperature-sensitive mutants or overexpression mutants need to be used to study the function of a gene, rather than a knockout mutant?

6. What are the features of phospholipids that allow them to form biological membranes? What joins phospholipids in making the outer membrane of Gram-negative bacteria?

7. Briefly, how do capsules help bacterial cells survive?

8. Using a drawing if you wish, differentiate between pili, fimbriae, and flagella.

9. What is the main difference between a secretion system and an efflux pump system, since both types of system are involved in transport out of the cell?

10. Contrast the difference in resolution of proteins on a 1D gel that resolves only by molecular weight and a 2D gel that resolves by isoelectric point and then by molecular weight.

11. Of the seven secretion systems, all except the Type 7 Secretion System involve both the outer and inner membranes. Why is the T7SS the exception and how else does it differ from the others?

12. How do efflux pump systems contribute to antibiotic resistance? Is this trait acquired or an innate feature of species that express efflux pumps?

Discussion topics

These topics are presented for discussion in study groups, as part of class discussions, or on your own. These questions go beyond what is directly covered in this part of the book. Use the research literature and other reading to explore these topics in more depth. Tips to help prepare for topic discussions are available on-line.

13. Create a table listing in each row the secretion systems thus far identified in bacteria. Make columns across the components within the system, the location of the system, whether it is ATP-dependent or not, and what the secretion system transports. Fill in the table with information on example secretion systems that have been well researched.

14. Create a table listing the different types of efflux pump systems in different rows. In each column, specify the membranes that the components span, whether the efflux pump system is ATP-dependent, and the substrates that the efflux pump transports. Include for each, examples of well researched efflux pump systems.

15. Investigate the ways in which MALDI-TOF-MS is being used diagnostically and in investigation of antibiotic resistance in bacterial populations. Discuss its use with regards to a particular bacterial species, by a certain hospital, or in addressing a specific health concern.

Further reading

Cellular structural components not encoded by genes but produced by enzymes

Rowlett VW, Mallampalli VKPS, Karlstaedt A, Dowhan W, Taegtmeyer H, Margolin W, Vitvac H. Impact of membrane phospholipid alteration in *Escherichia coli* on cell function and bacterial stress adaptation. *J Bacteriol*. 2017; 199(13): e00849–15.

Wang X, Quinn PJ. Lipopolysaccharide biosynthetic pathway and structure modification. *Prog Lipid Res*. 2010; 49: 97–207.

Bacterial secretion systems move proteins across the bacterial membranes

English G, Trunk K, Rao VA, Srikannathasan V, Hunter WN, Coulthurst SJ. New secreted toxins and immunity proteins encoded within the Type VI secretion system gene cluster of *Serratia marcescens*. *Mol Microbiol*. 2012; 86(4): 921–36.

Green ER, Mecsas J. Bacterial secretion systems—An overview. *Microbiol Spectr*. 2016; 4(1): 10.

All of the expressed proteins are the proteome

Callister SJ, McCue LA, Turse JE, Monroe ME, Auberry KJ, Smith RD, Adkins JN, Lipton MS. Comparative bacterial proteomics: Analysis of the core genome concept. *PLOS ONE*. 2008; 3(2): e1542.

Proteomics aids in identification of the core genome

Pérez-Llarena FJ, Bou G. Proteomics as a tool for studying bacterial virulence and antimicrobial resistance. *Front Microbiol*. 2016; 7: 410.

Part IV

Genetics, Genomics, and Bioinformatics

Humans have been doing genetic experiments for thousands of years through selective breeding of plants and animals to select for traits we desire in crops, flowers, pets, animals, and even each other. Although humans did not know the molecular basis for inheritance, it was known that traits and features passed from generation to generation. In the modern era, bacteria have been at the heart of genetics and genomics research. Genomics brings together genetics, molecular biology, biochemistry, statistics, computer science, and disciplines such as engineering and biotechnology in developing technology to study genomes. In this section, the field of bacterial genetics will be further explored and expanded upon. Genetics and genomics have made important contributions to our understanding of the bacterial world, which will be explored in this section, as will the subject of **bioinformatics**, which uses computer-based tools to analyze genomic sequence data. Information in this section will build upon previous topics, adding to the depth of understanding of genetics and genomics and introducing bioinformatics.

Genetics

This chapter will explain the field of bacterial genetics, starting from the terms and concepts particular to the discipline. Genetics is the foundation from which larger investigations, such as those in genomics, can be explored. Also presented are the origins of our understanding of genetics. Some of the key important insights gained by humans into genetics are highlighted here, demonstrating how we have arrived at our current state of understanding in the field. This chapter will begin to explore the genetic tools and techniques that have been used in the laboratory in the past and are still used today to unravel the information contained within DNA. Details on how these technologies work, will be addressed in Chapter 18.

Terms and conventions in the field of bacterial genetics are straightforward

By now, terms like gene, genetics, genotype, phenotype, genome, and chromosome should be quite familiar. Unlike disciplines such as immunology, bacterial genetics and genomics do not have a lot of abbreviations to try to remember. DNA and RNA are relatively commonplace abbreviations, which often do not need definition. The abbreviation CDS has been used in previous chapters and refers specifically to the coding sequence from the start codon to the stop codon within an open reading frame, or ORF (see Chapter 2). A gene is a CDS, but not all CDSs are genes. This is because a sequence with a start codon and a stop codon that has been identified in sequence data is only speculated to be encoding a protein, based on the sequence evidence. To be a gene, that CDS has to have been shown to actually generate a protein via transcription and translation. CDS is therefore a designation assigned from computational analysis of genetic data, whereas genes are defined by laboratory experimentation. This nuance is often overlooked, with sequences being called genes without evidence to support the designation and CDSs referred to as ORFs, when an open reading frame is actually defined as a region between two stop codons (see Chapter 2).

In bacterial genetics, the name of the gene matches the name of the protein. The gene *dnaA*, for example, encodes the protein DnaA (**Figure 10.1**). For the most part, bacterial genes are named with a three letter code that may correspond to a function or assumed function of the gene. This is followed by a capitalized letter. Different letters can help differentiate genes that are all involved in the same process. Gene names are always in italics and always start with a lower case letter. Protein names, however, are not italicized and the first letter is capitalized. In this way, the reader can differentiate between the gene and the protein. Care should be taken when writing to ensure which form is used and whether the topic actually refers to the gene or its encoded protein.

Figure 10.1 Nomenclature for genes and proteins. The DNA includes many genes, which are identified with names that reflect their function, such as these three genes involved in multiple antibiotic resistance, *marR*, *marA*, and *marB*. Note the differences in how the gene names are written (*marR*, *marA*, and *marB*) versus how the protein names are written (MarR, MarA, MarB). In this way, it is possible to recognize whether a gene or protein are being referred to in writing. These names are based on the functions of the genes, so here *mar* comes from *multiple antibiotic resistance*.

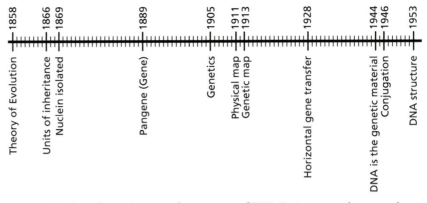

Figure 10.2 **Timeline of genetics up to the structure of DNA.** Key important dates over the century from Darwin's Theory of Evolution to Watson and Crick's structure of DNA.

Humans have understood about traits and inheritance long before the term genetics

Evidenced in the food in our markets and the pets with whom we share our homes, humans have had an understanding of the inheritance of traits for several thousand years. Modern dog breeds are very different from their ancestral wild relatives. Over time, through breeding practices where favorable traits were selected, humans were able to alter the genetic make-up of these animals. The same is true of the crops that are grown for our food, with humans having selectively bred and planted those that were high producers, resistant to stresses, and tasted best. Although we did not call these selective breeding practices genetics, they were the earliest forms of genetically modified organisms, with humans directly influencing the genetic make-up of another species.

It was not until 1858, that Charles Darwin described evolution of species and natural selection of traits. His concepts made possible a better understanding of inheritance of traits and could be directly applied to not only phenomena seen in the natural world, but also to the changes that humans had been making for over 10,000 years to the species around them. Gregor Mendel's famous experiments that looked at the traits carried by pea plants and kept careful record and detailed drawings of crosses between plants with different traits, was published in 1866. Mendel's observations led to the concept of there being units of heredity that were responsible for traits and that these units are unchanging particles. It was Hugo de Vries in 1889, who named the units of inheritance pangenes, which later became genes. It wasn't until 1905, that William Bateson coined the term genetics (**Figure 10.2**).

DNA was ignored and believed to be too simple to be the genetic material of inheritance

Friedrich Miescher published in 1869 about his discovery of a weak acid present in pus recovered from hospital bandages. This weak acidic substance was DNA, although at the time it was called nuclein because it was isolated from the nucleus. DNA was a very simple molecule and at the time was not believed to be the genetic material.

Although it may be hard for us to understand today, it was known that genes code for protein and that genes are on the chromosome, but it was not known that it was the DNA that made up the gene. Early in the 1900s, Mendel's work was rediscovered and applied to other organisms, including *Drosophila melanogaster* (fruit flies) which became a model organism for genetic research due to the ease of cross-breeding, easy to identify phenotypes like eye color and wing structure, and short generation times. The concept of genes, Mendel's units of inheritance, became well established and genetic experiments were happening well before it was known that genes are made of DNA.

Indeed in 1911, Alfred Sturtevant created the first physical map of a chromosome, which was expanded to a genetic map of the chromosome in 1913, highlighting where key genes could be found. Genetic transfer by the Griffith experiment, previously discussed (see Chapter 1), demonstrated that traits from one bacterial strain could be transferred to another strain. These results showed that chromosomes were the genetic material, supporting the knowledge and understanding of the time. These experiments also showed that bacteria participate in horizontal gene transfer. Some bacterial species, such as the streptococci used in the Griffith experiments, are able to take up DNA from other bacterial cells and incorporate it into their own genome, thereby expressing traits from the acquired DNA.

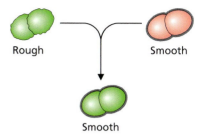

Figure 10.3 Horizontal gene transfer of the smooth trait in *S. pneumoniae*. Rough appearing bacteria (R strains) are not virulent (left, green cell). Smooth appearing bacteria (S strains) are virulent (right, pink cell), causing disease in humans. When rough and smooth strains are combined (arrows), the smooth trait can transfer to the rough strain (bottom, green cell), making them smooth. This demonstrates that the smooth trait can be transferred horizontally.

Bacterial genetics was the key to demonstrating the importance of DNA

As is highlighted throughout this book, several fundamental concepts in the fields of genetics and genomics have come about through research done in bacteria. One of the most important is the discovery made in the Avery, MacLeod, and McCarty experiment, which showed that DNA is the genetic material. These pivotal experiments demonstrated that the weak acidic substance first identified by Miescher from pus-filled hospital bandages was a critically important component of the cell and the key to genetics.

The rapid growth and short generation times of bacteria have been useful in studying their evolution and genetics, concepts that could be extrapolated to other organisms. Genetic passage of traits between bacteria, as observed by Griffith, Avery, MacLeod, and McCarty (see Chapter 1), has made important contributions to our understanding of genetics. In 1928, the Griffith experiment showed that transformation, the passage of genetic material from one bacterium to another, was possible. These experiments used *Streptococcus pneumoniae*, which caused important clinical infections at the time. Some *S. pneumoniae* strains had differences between strains that could be readily observed in the laboratory. Those agar cultures in which the bacterial colonies appeared rough were R strains, while S strains were smooth. The R strain *S. pneumoniae* did not cause disease and these non-virulent strains were able to be changed into virulent S strains in the course of the Griffith experiments (**Figure 10.3**).

There also seemed to be different virulent strains of *S. pneumoniae* that were causing bacterial pneumonia. Patients were able to make antibodies against one virulent S strain of *S. pneumoniae* but these antibodies were not necessarily able to recognize the antigens on other virulent S strains of *S. pneumoniae* in the laboratory (**Figure 10.4**). Oswald Avery studied the bacterial capsule that surrounded virulent S strains of *S. pneumoniae*. He established that the bacterial capsule is made of polysaccharide and discovered that different strains of *S. pneumoniae* had different and distinct polysaccharide capsules. Avery proposed that the capsules were antigenic and that this was the reason behind the differences in antibody responses seen in patients. The genes encoding the ability for *S. pneumoniae* to make distinct polysaccharide capsules are present in the DNA. Therefore, it is not only possible to transform a strain from having no capsule to expressing capsule, it is also possible to transform strains from one capsule to another.

Insights following the recognition of DNA as the genetic material led us to where we are today

The experiments of Griffith, Avery, MacLeod, and McCarty were early transformation experiments. Transformation is still used today to conduct genetic experiments on bacteria and for other research. There are, however, few bacteria that are naturally competent for transformation, meaning that there are not many species of bacteria that transfer traits in this way. Griffith was fortunate in his choice of bacterial species demonstrating transformation, a fortunate choice that Avery, MacLeod, and McCarty followed in their modifications of Griffith's experiments.

Later, other mechanisms for horizontal gene transfer were discovered, including the identification of conjugation in 1946. This involves the transfer of genetic material from one bacterial cell to another via a conjugation pilus and was described

Figure 10.4 Antigenicity of S strains of *S. pneumoniae*. A patient (top left) infected with an S strain of *S. pneumoniae* (green bacteria) makes antibodies specific to the antigens on the surface of that S strain (middle left). These antibodies cannot recognize other S strains, only the one that infected the patient (bottom left). Different patients infected with different S strains, produce different antibodies, each only able to recognize the S strain that infected the patient (right panel).

end of the DNA after it is cleaved. In the type 2 restriction modification systems there are separate restriction and modification enzymes. The restriction enzyme forms a **homodimer**, where two copies of a protein come together into one quaternary structure that is then capable of specifically binding to the DNA. This homodimer structure influences the nature of the recognition site, which is a palindromic sequence of four to eight nucleotides. Cleavage of the DNA occurs at this site, which is the key feature making these enzymes the most commonly used and commercially available restriction enzymes.

The first type 2 restriction enzyme that was identified and isolated was the *Hind*II enzyme from *Haemophilus influenzae*. Restriction enzyme nomenclature comes from the species in which it was first identified, making up the first three letters of the enzyme name, which is written in italics. The remaining letters identify the strain of the bacteria and then Roman numerals based on the order in which restriction enzymes were identified in this strain. So, the type 2 restriction enzyme *Eco*RI comes from *E. coli* strain R and was the first enzyme identified in this strain. *Hind*I is a type 1 restriction enzyme, with *Hind*II, the first type 2, discovered later.

Type 3 restriction enzymes are a complex of restriction and modification enzymes that cleave DNA a short distance from the recognition site. The recognition sequence for a type 3 restriction enzyme is actually two different sequences that are present in inverted orientation relative to one another.

Unlike the other types of restriction enzymes, which are blocked by methylation generated by the associated methylase, the type 4 restriction enzymes specifically cleave methylated DNA. Type 4 restriction modification systems are otherwise similar to type 2 systems. Some classify the enzymatic activity within a CRISPR system as a fifth restriction enzyme type; this will be explored in more detail in Chapter 18.

The genetics of bacteria was unraveled using conjugation

In 1957, Jacob and Wollman interrupted conjugation in the laboratory to determine the location of genes on the *E. coli* chromosome. Bacteria with the F plasmid integrated into the chromosome as Hfr (see Chapter 3) were used to transfer the bacterial chromosome from one *E. coli* cell to another. As the chromosome crosses from one bacterial cell to another during conjugation, the gene locations on the chromosome were mapped based on the time the bacteria are attached by the conjugation machinery. During conjugation, the conjugation pilus retracts, pulling the donor and recipient cells closer together. This enables this Type 4 Secretion System (see Chapter 9) to transfer the DNA between the bacterial cells. Interruption of the conjugation at different times would allow a different amount of the chromosome to transfer over to the recipient. Conjugation experiments were interrupted quite simply – using a standard blender to separate the bacterial cells. Because the F plasmid can integrate into different locations in the chromosome in different Hfr strains, repeated experiments would produce results around the chromosome identifying the relative locations of genes.

The conjugation mapping experiments relied upon **auxotrophy**, the requirement for particular nutrients in the media for growth. **Auxotrophs** require specific nutrients in the media to grow, therefore they can be easily identified through the use of media with different nutrients. Auxotrophies can differentiate bacteria that are otherwise identical, but which need different nutrients for growth. Genetic mapping experiments can track the locations of auxotrophy genes. The transfer of genes between cells in conjugation can reverse the auxotrophy, highlighting the genetic basis of the auxotrophic phenotype. In addition, genes conferring antibiotic resistance enable bacterial cells to grow on media containing the antibiotic. Timing this transfer of genes identifies the location of the gene on the chromosome. Through tracking of auxotrophy genes and antibiotic resistance genes, it was possible to generate a 100 minute map of *E. coli* that represented the complete conjugative transfer of the whole *E. coli* genome (**Figure 10.7**). These experiments were difficult; the connection between conjugating bacteria is very easy to break. Maintaining a connection for a few minutes is relatively easy, but maintaining connections for up to 100 minutes is technically challenging.

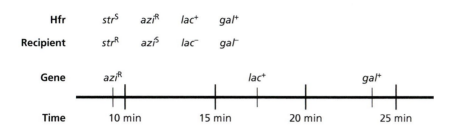

Figure 10.7 Conjugation mapping of *E. coli* using auxotrophies and antibiotic susceptibility. The Hfr strain is susceptible to streptomycin (*str*), resistant to azithromycin (*azi*), able to use lactose (*lac*), and able to use galactose (*gal*). The recipient strain is resistant to streptomycin, susceptible to azithromycin, and unable to use lactose or galactose. Conjugation is allowed to happen for a set time and then disrupted to stop conjugation. After conjugation, bacteria are cultured on media containing streptomycin, to select for the recipient cells. Only the recipient cells will grow, as the Hfr strain is susceptible to streptomycin. The time points at which these streptomycin-resistant recipient strains gain resistance to azithromycin and the ability to grow on lactose and on galactose produce a map of the genome with the relative distances of these genes indicated in the minutes of conjugation required to transfer the trait.

Through these timed conjugation experiments the relative positions of readily identifiable genes were determined. This produces a **genetic map** that gives the relative locations of genes and genetic features on the chromosome, such as auxotrophy genes, antibiotic resistance genes, and other genes of known function. In this way, the genetic features on the bacterial chromosome were first explored. While valuable, and certainly a cornerstone of understanding the *E. coli* genome for many years, conjugation does not happen in all bacterial species, therefore generation of these genetic maps was not possible for every species of interest.

Physical maps can be made for any bacterial species using restriction enzymes

A **physical map** does not have gene location information like a genetic map. Instead, the base pair distances between restriction enzyme recognition sites around the chromosome are determined. The genetic features on physical maps are the type 2 restriction enzyme recognition sequences. Physical maps are made by digesting the chromosome with restriction enzymes and separating the **restriction fragments** by electrophoresis. The pattern of fragments on a gel is used to identify the lengths of the fragments. When several different restriction enzymes are used, singly and in **double digests** using two enzymes at the same time, the locations of these different recognition sites in relation to one another can be determined by determining the size of the restriction fragments and analyzing how they would go together. Because restriction enzymes are used to generate the map, these are also referred to as **restriction maps**. Restriction maps can be made for chromosomes and also for plasmids. Plasmid restriction maps are useful when planning to digest the plasmid with a restriction enzyme for the purposes of cloning (**Figure 10.8**).

In a physical map, the genes and other genetic features on the chromosome or plasmid are not identified. It is possible to add genetic map information onto a physical map by experimentally determining the genes within a restriction fragment. A physical map can be used, when overlaid with a genetic map, to identify

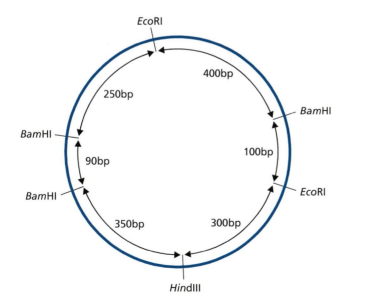

Figure 10.8 A physical map of a plasmid. Physical maps use restriction enzymes to generate DNA fragments, which are used to plot the distances between restriction enzyme recognition sites. Using several different enzymes can create a detailed map of the locations of the various enzyme recognition sites. Shown here are the locations of the restriction enzyme recognition sites for *Eco*RI, *Bam*HI, and *Hind*III.

restriction fragments that contain genes of interest and then use this information to isolate the fragment of DNA containing the gene. Because the restriction enzyme recognition sites are specific known sequences and the distances between them are determined in the physical map, this resource can be used to help in assembly of genome sequences and then to check that the assembly is accurate. The availability of a physical map was deemed to be important in the selection of strains for the first genome sequences. Although new physical maps using restriction enzymes are often not determined today, the methods behind their creation can still be used to verify genomic data. Many of the known functions of genes were determined through experiments that developed genetic and physical maps and experiments that used these data.

Experimentation reveals whether a CDS is a gene and what its function may be

While in the past mapping and molecular biology techniques were used to identify genes, today the abundance of sequencing data means that potential genes tend to be identified from DNA sequence, rather than experimental data. Although potential genes can be identified as CDSs through analysis of sequence data, only laboratory experiments can truly reveal the expression of the CDS as a gene and its function. Although there are many examples in a bacterial genome of regions where there is a start codon, a segment of uninterrupted DNA, and then a stop codon, identification from the sequence is based on sequence information, not on expression of that sequence as a protein. Even in CDSs with homology to known genes in other species, the function of the potential protein is only assumed to be the same as that of the homologous gene, it is not known to be the same.

Experiments in which a CDS is deleted or mutationally inactivated can demonstrate that this sequence includes a gene and can reveal its function. If, in the absence of the CDS sequence a phenotypic difference is observed in the bacteria, the CDS can then be designated as a gene. If the phenotype gives a clear indication of the function of the gene, this is used to name the gene, using the conventions discussed earlier in this chapter. CDSs should not be given gene names when identified from sequence data alone. For this reason, CDSs are given other designations, often based on their relative locations around the chromosome, so that they can be identified and discussed in the literature prior to their status as a gene being known.

Library generation and library screening can identify genes and their functions

Transposon mutagenesis uses the insertion of transposons (see Chapter 3) into a bacterial chromosome to create a **library** of nearly identical bacterial **clones** where each has one transposon somewhere in the genome, disrupting the chromosomal sequence. Transposons used for the construction of libraries are specifically chosen to have antibiotic resistance markers that can be readily selected, to randomly insert in the genome, and to be small in size, sometimes being referred to as mini-transposons. The disruption of the genome is not targeted, but relies on the promiscuous integration of the transposon to semi-randomly generate mutations across the chromosome, causing **random mutagenesis**. The process is semi-random due to the nature of the target sites for the particular transposon used. The library is a collection of strains that can be used to look for changes in phenotype based on the different insertions of the transposon within the library.

It is possible to use **selection**, to select library clones with a particular phenotype through the creation of a specific media on which only those clones will survive. Auxotrophs and antibiotic resistance markers can easily be identified in this way. Library clones that grow can then be analyzed in more depth to see where the transposon has inserted in that particular bacteria. The transposon may have interrupted a gene directly involved in the selected phenotype. It may also be that the transposon has inserted into a promoter region that is important for expression of the phenotype or into a regulatory protein gene needed for expression of the phenotype.

Libraries can also be interrogated for specific phenotypes, using **screening**. Unlike selection, all of the bacteria in the library are grown and then each individual clone is experimentally investigated for the presence of the phenotype, using a variety of molecular biology methods. This is a labor intensive process because each library clone must be investigated individually for the phenotype of interest. The larger the library, and therefore the more extensive the transposon mutagenesis of the genome, the more experiments will be needed to identify the clone of interest. Like with selection, once the clone or clones with the desired phenotype are identified, the location of the transposon is then determined, using PCR or sequencing, to identify genes involved.

Random mutagenesis identifies genes that have non-essential functions

A transposon library is a group of clones that are all identical except for the location of insertion of the transposon. Each clone has a transposon somewhere in the chromosome, interrupting the DNA sequence. If this mutation is detrimental to the bacterial cell, it will not grow and will not be part of the transposon library. This means that only non-essential genes will be disrupted by the transposon.

Locations in the chromosome that do not have transposon insertions in the transposon library will correspond to essential genes. When the locations of the various transposon insertions across the chromosome are assessed, the lack of any transposon insertions in a region suggests that there is an essential gene there, or another genetic feature that is essential, such as a promoter region driving essential genes or a ncRNA that the bacterial cell needs for growth.

If the growth conditions are changed for the culture used to grow the transposon library and some mutants are lost, the insertion locations of the transposons in the lost clones may be involved in growth in the new condition. By identifying the lost clones from the original culture conditions it is possible to identify the insertion location and associate the phenotype with the genetic feature.

A specialized form of transposon library is called **signature-tagged mutagenesis** (**STM**). When the transposon library is made, specialized transposons are used that contain specific sequences, genetically engineered tags, within the transposon. Using molecular techniques such as PCR, the specific clones within the library can be identified. In experiments using STM, the transposon mutants are pooled and subjected to different growth conditions to see which can and which cannot survive in the different experimental growth conditions (**Figure 10.9**). Often signature-tagged mutagenesis has been applied to growth in laboratory *in vitro* conditions versus *in vivo* conditions in a host organism to identify genes required for growth *in vivo*. The signature tags enable the identification of clones more readily in the library. High throughput, inexpensive sequencing means identified mutants from either standard transposon libraries or STM can be whole genome sequenced to identify the location of transposon insertion.

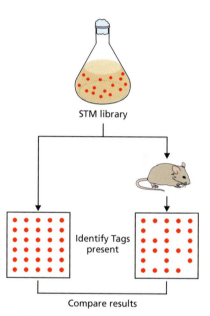

Figure 10.9 Signature-tagged mutagenesis to identify genes needed *in vivo*. A signature-tagged library is grown in laboratory conditions. A portion of this culture is injected into mice (or another *in vivo* model system) to assess which library clones are lost *in vivo*. These clones will have disrupted genes that are essential for survival *in vivo*. By virtue of the tags, the clones from the laboratory culture and from the *in vivo* model can be assessed and those results compared to identify missing clones from the *in vivo* conditions.

The functions of genes can be determined using knockout technologies

To find the phenotype for a gene, a specific **knockout** of that gene may reveal the function. Disruption of the gene, the knockout, should result in a loss of function of that gene product. Libraries are particularly useful when the gene for a specific phenotype is sought, or when the functions of several genes need to be investigated. The identified loss of function should relate to the gene, although it should be noted that it is not always that simple.

In the same way that a transposon interrupts a gene sequence, a knockout mutation specifically gets rid of a gene. This can be achieved by specifically inserting a different sequence into the gene, disrupting the sequence of the desired gene and rendering it unable to encode the protein. Unlike transposon mutagenesis, the gene is not disrupted by random process, but rather in a carefully designed experiment. Disruption of the gene sequence is often achieved with an antibiotic

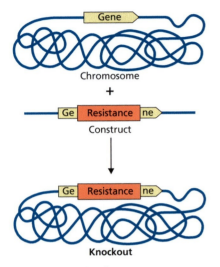

Figure 10.10 Knockout mutant generation. To knockout a gene of interest on the chromosome (top), a construct is generated, either on a plasmid or using PCR. A fragment of DNA containing the gene is created that has an antibiotic resistance cassette inserted into the gene, disrupting the sequence of the gene and thus preventing its function (middle). When this fragment recombines with the chromosome, the homology between the portion of the gene sequence in the construct and that in the chromosome means the construct ends up in the chromosome, generating the knockout of the gene.

resistance cassette, which means that clones carrying the insertion in the gene can be selected for on media containing the antibiotic. Homologous recombination occurs between the chromosome and a fragment of engineered DNA containing the antibiotic resistance cassette flanked by sequences homologous to the chromosome. The flanking sequences are chosen so that recombination deletes the gene of interest. Bacterial cells that have not incorporated the antibiotic cassette into their chromosome will not be resistant to the antibiotic and will fail to grow on the plate, leaving only the mutants to grow. This knockout mutant can be used in experiments to try to determine the function of the disrupted gene.

Advances in molecular biology and molecular cloning mean that genes can be put into a plasmid and manipulated there. This **construct** enables the manipulation of the genetic sequence of interest. Once inserted into the plasmid, it is possible to insertionally inactivate the gene, delete all of the gene sequence, or delete part of the gene for the purposes of creating a knockout. A construct can also be generated using PCR techniques, which will be explored in detail in Chapter 18. Recombination between the genetically modified sequence and the bacterial chromosome will bring the changes to the gene into the bacterial genome, generating a knockout mutant (**Figure 10.10**). Molecular cloning techniques will be discussed in more detail in Chapter 18.

Knockout a gene and complement it back to check the phenotype is caused by the knocked out gene

Since the observed phenotype in a knockout mutant could be due to something else having changed in the genome, it is important to add back the gene and show a return of function. This process is called **complementation** and is often achieved by providing an intact copy of the gene on a plasmid (**Figure 10.11**), that is introduced into the bacterial cell. During the process of generating the knockout, a mutation could have appeared elsewhere in the genome and this could cause the phenotype observed and presumed to be attributable to the knocked out gene. Complementation provides additional supporting evidence that the phenotype associated with the knockout has actually been caused by the disrupted sequence of the knockout.

In the era of inexpensive genome sequencing, it is also possible to check for mutations and deletions of the genome that may have arisen during creation of the knockout mutant through whole genome sequencing. Invariably there are changes seen in the whole genome sequencing data. Even if two colonies from the same culture plate are sequenced, there can be differences due to errors introduced during replication and sequencing errors from the sequencing technology. Therefore, sequence data from knockout mutants have to be carefully assessed to see if there are sequencing errors or random mutations that do not impact the phenotype.

Through either targeted knockouts or random mutagenesis, like transposon libraries, many gene functions are identified. However, these techniques cannot be used on genes that are essential to bacteria. Deletion or disruption of a gene that is essential for bacterial growth and survival will result in dead bacteria that cannot be investigated in the laboratory. For essential genes other strategies are needed.

Figure 10.11 Complementation of a knockout mutant. To add to the evidence that a knockout mutant is responsible for an observed phenotype, the gene is added back into the bacterial cell, causing complementation. Often this is achieved by putting a functional copy of the gene on a plasmid and introducing it into the bacterial cell.

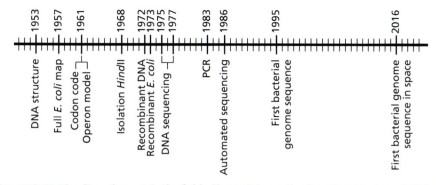

Figure 10.12 Timeline of events in the field of bacterial genetics from the structure of DNA to the present.

Bacterial research has helped shape the field of genetics

Several key events in recent decades involving bacteria have occurred that contribute to the general field of genetics. In the 1950s the DNA code was determined, and the principles of transcription, translation, replication, and operons were developed, as discussed briefly here and in earlier chapters. In the 1970s, key enzymes were identified, including restriction enzymes and reverse transcriptase, which are routinely used in molecular biology and have contributed to our understanding of genomes and transcriptomes. Recombinant DNA was first generated by Paul Berg in 1972, with 1973 seeing the creation of recombinant *E. coli* expressing non-*E. coli* genes, including a gene from *Xenopus laevis*, a toad. The first gene sequenced was from bacteriophage MS2, with both Sanger sequencing and Maxim and Gilbert sequencing being applied to bacterial sequences soon thereafter. Building upon earlier work and understanding of replication in the bacterial cell, Kary Mullis introduced the world in 1983 to the **polymerase chain reaction** (**PCR**) enabling researchers to copy specific sequences of DNA in the laboratory. PCR will be explored in depth in Chapter 18. The first automated sequencing machine was available in 1986, an enhancement of Sanger sequencing methodology that enabled the first bacterial genome sequence in 1995 of *H. influenzae*. Indeed, advances in technologies are often first applied to bacteria, due to the small size of their genomes and relative ease for genetic manipulation. As such, in 2016, *E. coli* was sequenced in space using a nanopore MinION system by astronaut Kathleen Rubins (**Figure 10.12**).

Key points

- Bacterial genetics is the study of how the nucleic acids within the bacterial cell contribute to its characteristics.

- Bacterial gene names are always in italics and the proteins they encode have the same name, but capitalized, making it easy to identify the gene and product in writing.

- Many of the advances in understanding the wider field of genetics have been made through experiments in bacteria, including identifying DNA as the genetic material.

- *E. coli* is often used in molecular biology for molecular cloning and expression of recombinant proteins, making it an important microorganism.

- The specificity of restriction enzymes in their cleavage of DNA has made possible the generation of physical maps of bacterial chromosomes and the creation of recombinant DNA, key advances in genetics.

- Timed conjugation experiments created genetic maps of bacterial genomes before it was possible to sequence the genome.

- From sequence data, we can predict where a gene might be and what it might do, but only through experimentation can we demonstrate which sequences are truly genes and what they do.

- Transposon mutagenesis, and variants of this method, randomly generates mutants that can contribute to understanding the functions of genes.

- Targeted gene knockouts and complementation can be used to investigate the function of a specific gene of interest.

Terms, questions, and discussions

Key terms

Define the following terms introduced in this chapter. Check your answers using the definitions in the Glossary. These terms are also available as Flashcards on-line.

Auxotroph	*In vitro*	Recombinant protein
Bioinformatics	*In vivo*	Restriction enzymes
Blunt ends	Knockout	Screening
Clone	Molecular cloning	Selection
Complementation	PCR	Sticky ends
Construct	Recombinant DNA	Transposon mutagenesis

Self-study questions

Answer each question using 50–100 words or a table or labeled diagram. Advice on where to find answers to these questions is available on-line.

1. Describe the distinction between a gene, a CDS, and an ORF.

2. How is it possible to distinguish in writing between a gene name and a protein name? What is the general convention for naming bacterial genes and the proteins that they encode?

3. Early experiments in genetics used the fruit fly, *Drosophila melanogaster*, to study phenotypes. Why was this model organism used and continues to be used today?

4. What mechanism was involved in the horizontal gene transfer that occurred in the experiments by Griffith, Avery, MacLeod, and McCarty with *S. pneumoniae*? What phenotype was horizontally acquired?

5. What are bacterial capsules made from and are all capsules within a species the same? How do capsules impact the immune response to bacteria?

6. What does a restriction endonuclease do to a piece of DNA containing its recognition sequence? Draw the different outcomes for the DNA.

7. Compare the differences between a genetic map and a physical map.

8. What can be done experimentally to demonstrate the function of a CDS?

9. What is the counterpart experiment to the generation of a knockout, which helps to verify the phenotype of the loss of function seen in the knockout? How would these experiments show the function of an unknown gene?

10. What is the difference between selection and screening? Which do you think is quicker or easier to do?

11. What sort of genes cannot be investigated using knockout mutants and transposon libraries? Why can they not be investigated in this way? Can essential genes be identified from library data?

12. When was recombinant DNA first generated? Which species were involved in these experiments?

Discussion topics

These topics are presented for discussion in study groups, as part of class discussions, or on your own. These questions go beyond what is directly covered in this part of the book. Use the research literature and other reading to explore these topics in more depth. Tips to help prepare for topic discussions are available on-line.

13. Just as *Drosophila melanogaster* was chosen as a model organism for the study of eukaryotic genetics, *Escherichia coli* became the model organism for the study of bacterial genetics. Discuss the reasons why this was chosen as a readily available bacterial species for laboratories to investigate and to manipulate as molecular biology developed. How have discoveries made in *E. coli* been applicable to a broad range of other bacterial species, both Gram-negative and Gram-positive?

14. Identify a recent research article that has explored the function of a bacterial gene. Discuss how the researchers determined the function of the gene, including the genetic experiments conducted and the tests run to demonstrate the function. Was a knockout mutant made and loss of function shown? What were the controls in the experiment? What more needs to be done in future research?

15. Several important advances in the wider field of genetics and genomics have been based on work in bacteria. Within the last year, what vital research discoveries have been made due to investigations in bacterial species? Discuss the implications of these discoveries for the wider field of genetics and genomics and potential applications for the future.

Further reading

Bacterial genetics was the key to demonstrating the importance of DNA

Avery OT, Goebel WF. Chemoimmunological studies on the soluble specific substance of pneumococcus I: The isolation and properties of the acetyl polysaccharide of pneumococcus type I. *J Exp Med*. 1933; 58: 731–55.

McCarty M. Discovering genes are made of DNA. *Nature*. 2003; 421: 406.

Steinman RM, Moberg CL. A triple tribute to the experiment that transformed biology. *J Exp Med*. 1994; 179: 379–84.

Insights following the recognition of DNA as the genetic material led us to where we are today

Crick FH, Barnett L, Brenner S, Watts-Tobin RJ. General nature of the genetic code for proteins. *Nature*. 1961; 192: 1227–32.

The identification and isolation of restriction enzymes is important for genetics research

Pingoud A, Wilson GG, Wende W. Type II restriction endonucleases – A historical perspective and more. *Nucleic Acids Res*. 2014 Jul; 42(12): 7489–527.

Library generation and library screening can identify genes and their functions

Lin T, Troy EB, Hu LT-Gao L, Norris SJ. Transposon mutagenesis as an approach to improved understanding of *Borrelia* pathogenesis and biology. *Front Cell Infect Microbiol*. 2014 May 20; 4: 63.

Random mutagenesis identifies genes that have non-essential functions

Cummins J, Gahan CG. Signature tagged mutagenesis in the functional genetic analysis of gastrointestinal pathogens. *Gut Microbes*. 2012 Mar–Apr; 3(2): 93–103.

Bacterial research has helped shape the field of genetics

Castro-Wallace SL, Chiu CY, John KK, Stahl SE, Rubins KH, McIntyre ABR *et al*. Nanopore DNA sequencing and genome assembly on the international space station. *Sci Rep*. 2017 Dec 21; 7(1): 18022.

Genomics

Bacterial genomics, in which the whole-genome sequence data is generated, interrogated, and explored, has only been possible since 1995. In that year, the first bacterial complete whole-genome sequences became available and the era of genomics truly began. Previous genomic endeavors had to rely upon incomplete information, gained from techniques such as mapping and mutagenesis, as described in Chapter 10, and short targeted sequencing. Since the first bacterial genome sequences were completed, many thousands of bacterial samples have been sequenced, leading to a new understanding of the genes and sequence features within bacterial species and how they have diverged and evolved over time. This recent history of genomics will be explored in this chapter, highlighting key advances in technologies, discoveries that have been made, and how bacterial genomics has impacted our understanding of bacteria.

Automation of Sanger sequencing launched the era of bacterial genome sequencing

Sanger sequencing and Maxim and Gilbert sequencing, developed in the mid-1970s, allowed the specific base sequences of DNA to be determined using biological or chemical techniques, respectively. These methods were technically challenging, therefore, the automation of Sanger sequencing made it possible to produce more sequence data more efficiently than ever before. This brief overview here will put into context the advances to Sanger sequencing that have led to genome sequencing.

Although Maxim and Gilbert sequencing was initially more popular, due to its chemical basis of sequence determination, ultimately Sanger sequencing became dominant, with its **sequencing by synthesis** approach. Sequencing by synthesis reproduces the process of replication, taking a single strand of DNA and using polymerase to add nucleotides. Through monitoring the addition of the four nucleotides, the sequence is able to be determined. Sanger sequencing initially used radioactively labeled nucleotides incorporated into the synthesized second strand of DNA to produce fragments and resolved these on large polyacrylamide slab gels. The radiolabeling of the four different nucleotides required four reactions to be set up and then resolved on four different lanes of the slab gel. After the gel was run, it was exposed to x-ray film and the radiation revealed the location of labeled nucleotides. To obtain the DNA sequence, the bands on the x-ray film across the four lanes were read by eye to reveal the order of the four nucleotides in the sequence (**Figure 11.1**).

To automate Sanger sequencing, the radiolabeling had to be replaced with something more easily detected by a computer-driven automated system, rather than relying on human interpretation from bands on a gel. **Automated Sanger sequencing** incorporated fluorescently labeled nucleotides. This change resulted in three main benefits. Using fluorescently labeled nucleotides, the bands on the polyacrylamide gel are detectable by a sensor and recorded automatically by a computer, rather than needing to be read and recorded by a human. Also, by using different fluorescences for each nucleotide, the four sequencing reactions needed in Sanger sequencing are able to be combined into one reaction in one tube, reducing set-up time for the sequencing reaction. Finally, since there is one sequencing reaction, containing information on all four of the nucleotides, the number of lanes needed for each stretch of sequence on the polyacrylamide slab gel is reduced from four lanes to just

Figure 11.1 Sanger sequencing. The original Sanger sequencing method required radioactive nucleotides (indicated with a *), one for each base within DNA. These were used in four separate reactions to label DNA (top). Each reaction was run in a single lane of a polyacrylamide gel (top middle), which was then exposed to x-ray film to reveal the radioactive bands, each representing a base in the sequence (bottom middle). This x-ray film was read by eye, from bottom to top, to determine the sequence of the DNA (bottom).

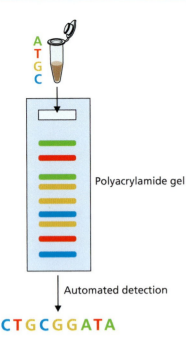

Figure 11.2 Sanger sequencing with fluorescent nucleotides. Rather than using four separate radioactive nucleotides in the sequencing reactions, use of nucleotides each labeled with a different type of fluorescence reduced both the number of reactions (top) and the number of lanes on the polyacrylamide gel (middle). Fluorescence also enabled automated detection (bottom), reducing human error in reading bands on an x-ray film as previously (see Figure 11.1).

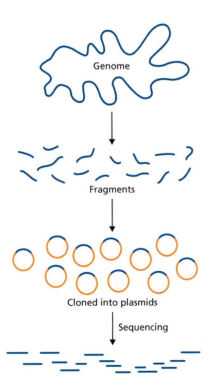

one lane. More efficient use of the slab gel means that four times more sequence data can be gathered for each gel poured and run. This dramatically reduced preparation time for both the reaction and the gel, since fewer reactions and fewer gel lanes were needed to achieve the same amount of sequence data (**Figure 11.2**).

This stepwise advance in sequencing methodology still required researchers to pour large awkward polyacrylamide slab gels, therefore, these were replaced in the early 1990s with capillary gels. By containing each gel lane within a glass tube, the process was simplified in a way that could be completely automated. From this stage, other advances were based on increasing the capacity of the sequencer, such that more capillaries could be run at the same time and such that more sequence information could be gained from each capillary. These developments and the potential seen in early Sanger sequencing launched the sequencing of bacterial genomes. As groups embarked on large-scale sequencing efforts, new insights into the processes involved in sequencing led to these higher throughput methods, automating Sanger sequencing and transitioning to capillary gels.

The first genome sequence of a free-living organism was bacterial

The first genomes that were sequenced were the relatively small genomes of bacteriophages. Scaling up to a bacterial genome sequence was a major undertaking, going from a few hundred bases to several thousand. The first bacterial species for which genome sequencing was undertaken was the model microorganism *Escherichia coli*, which has 4.6 megabases in its genome. This project took an ordered and methodical approach to bacterial genome sequencing. The *E. coli* chromosome was physically and genetically mapped previously, providing a reference against which sequence data could be placed. The genome was then ordered into 40,000 base segments that overlapped one another. Each 40,000 base segment became the focus of sequencing efforts. This facilitated the assembly of the 200–800 bases of sequence data from each Sanger sequencing reaction into the correct order in each segment. The overlapping segments were then connected to each other to form the complete *E. coli* genome. The first 1.9 megabases of sequence data from this project was deposited into GenBank in 1992 and the project was finished in 1997.

E. coli was the first genome-sequencing project initiated, but it was not, however, the first to be finished. The first bacterial genome sequence, completed in 1995, was *Haemophilus influenzae*. This involved determining the sequence of the 1,800,000 bases of the *H. influenzae* genome using a new **shotgun sequencing** approach and took about a year to finish. In this method, the genome is randomly divided, cloned into a randomized library, and sequenced. Because the sequencing data is then produced in a random order, assembly of the genomic sequence relies heavily on computation and upon there being enough overlapping sequence fragments to determine where each 200–800 bases of sequence data join to one another (**Figure 11.3**). Although shotgun sequencing is more computationally intense, it is less laborious than making ordered libraries, reducing both cost and time. In the end, about 24,000 fragments of DNA sequence data became the *H. influenzae* genome sequence.

In the early years of bacterial genomics, most genome-sequencing projects were completed to one contiguous sequence, with all of the generated sequence fragments joined up as they would be within the bacterial cell. The goal in each case was to fully sequence all of the bases in the genome and fully annotate each coding sequence of the bacterial genome. There were, however, many limitations in the early days of bacterial genome sequencing. The processes of sequencing used for the first bacterial genomes were laborious and expensive, which in turn drove progress toward automating and increasing throughput of the Sanger sequencing

Figure 11.3 Shotgun sequencing. This methodology, applied to sequencing whole-bacterial genomes, involves first randomly dividing the genome into fragments that are cloned into a library. The order to the fragments as they were in the genome is lost. As the library is sequenced, it produces overlapping sequence data that is joined computationally to generate contiguous sequence data.

methodology. Another limitation discovered during the course of the sequencing projects was that the libraries generated to facilitate all sequencing projects did not contain the entire genome. Regions of DNA that contained genes that were toxic to *E. coli* when these regions were cloned into plasmids and put into *E. coli* resulted in gaps in the sequence data. Because plasmids with these regions were toxic, no *E. coli* containing them were able to grow and therefore these regions could not be recovered in the resulting sequence data. Cloning into plasmids enables sequencing to use primers designed to anneal to the plasmid and sequence the inserted fragment. However, reliance on the propagation of the fragments within plasmids *in vivo* risks the loss of sequence data due to viability of clones. In order to fill in these gaps and complete the genome sequence, additional laboratory work and additional automated Sanger sequencing, targeting the missing regions, was required.

Early genome sequencing of bacteria provided opportunities for new insight and innovation

The availability of a bacterial genome sequence enabled researchers to explore the potential within the DNA, identifying features that might be genes and hypothesize as to their functions. As more whole-genome sequences were determined, new types of analyses could be done. **Comparative genomics**, in which two or more genome sequences are compared to one another, meant that it was possible to analyze similar sequences and identify **regions of difference**, where genomes contain different genetic features that are not in common between them (**Figure 11.4**).

The first comparative genomics study was published in 1996, evaluating the differences and similarities between two genomes from different *Mycoplasma* species: *Mycoplasma genitalium* and *Mycoplasma pneumoniae*. Due to the small size of *Mycoplasma* spp. genomes, *M. genitalium*, with a 580 kb genome, was chosen as the second target of the new shotgun sequencing technique in 1995. In the following year, the 816 kb genome of *M. pneumoniae* was also sequenced. This provided a unique opportunity to conduct comparative genomics, for the first time comparing two complete genome sequences of species that were within the same genus and therefore closely related. All of the CDSs in *M. genitalium* were found to be within the *M. pneumoniae* genome sequence data, although the order of the CDSs was different, indicating chromosomal rearrangements between the two organisms. Additional CDSs in *M. pneumoniae* included new unique CDSs as well as some duplications of CDSs present as single copies in *M. genitalium*. This first comparative genomics study showed that the genome sequence data was able to reveal the depth of variation between species and enabled researchers to suggest how changes may have occurred during evolution.

Following on from this, a comparative genomics project comparing two strains of the same species provided even greater insight into bacterial genomic variation. *Helicobacter pylori* became the first comparative genomics study to be published within a single species. This analysis revealed that each of the two *H. pylori* strains contained strain-specific CDSs, many of which were present together in regions of difference between the strains. The value of these comparative genomics results in understanding strain diversity and bacterial genome evolution, coupled with growth of sequencing technology and reduction of sequencing costs, encouraged investigators to sequence multiple species of the same genus and multiple strains of the same species in the years since.

Bacterial genome sequencing required and fueled innovation

In 1997, the first genome sequences of the model bacterial organisms *E. coli* and *Bacillus subtilis* were completed, adding to the growing list of bacterial genome sequence data. The growth in size of sequence data from database entries of a few kilobases to a few megabases proved to be a challenge when it came to interrogating this data in available software and when it came to storage and retrieval of the data for analysis. The data therefore drove innovation into new ways in which the vast

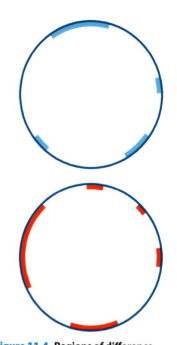

Figure 11.4 Regions of difference between bacterial genomes. Comparative genomics, evaluating differences and similarities between two genome sequences (top and bottom), shows that there are some regions in one genome that are not in the other (blue), and vice versa (red).

(in comparison to computer storage at the time) and growing amount of sequence data could be stored and the ways in which it could be analyzed. Without these innovations, investigations into the data would be very limited. These sequencing efforts into bacterial genomes were occurring at the time when the Human Genome Project was being pursued by a collection of sequencing centers around the world. Using the relatively smaller bacterial genomes, developers were able to test the ability of new software and data retrieval methods to enable them to ultimately analyze the larger human genome sequence.

By the year 2000, bacterial genome sequencing and comparative genomics brought about another innovation, **reverse vaccinology**. Using the genome sequence data, analysis was done to identify all of the potential targets for a new *Neisseria meningitidis* vaccine. It was known that the targets for a vaccine needed to be expressed on the surface of the bacteria and that they needed to be unique to bacteria. Theoretically, all of the proteins expressed on the surface of bacteria are encoded in the genome sequence, it was just a matter of doing the correct analyses to identify them. Following extensive sequence analysis and years of laboratory research, new vaccines were developed using this reverse vaccinology approach. Reverse vaccinology will be discussed in more depth in Chapter 20.

The emergence of next-generation sequencing technologies greatly increased sequence data

Early sequencing efforts using Sanger and automated Sanger methods demonstrated that bacterial genome sequencing could be achieved and would reveal important information about the bacterial species being studied. However, limitations to the application of these sequencing methods to large numbers of bacterial genomes remained, including the time, effort, and costs involved. High-throughput **next-generation sequencing** technologies dramatically reduced the time, effort, and cost per genome.

Several high-throughput technologies emerged in the mid- to late-2000s, most of which were based on similar principles. **Sequencing platforms**, the technologies and equipment for sequencing, by Roche (developed by 454), Illumina (developed by Solexa), and Life Technologies (developed by Ion Torrent) all used modifications of the sequencing by synthesis method pioneered by Sanger sequencing. Next-generation sequencing takes an approach that is massively parallel, with many individual sequencing reactions happening and being recorded within the sequencing equipment all at the same time. For each of these three next-generation sequencing technologies, the DNA is fragmented, similarly to shotgun sequencing. However, for next-generation sequencing these fragments are not cloned into plasmids. This addressed the previous methodology's issues with gaps in the final sequencing data due to non-viable clones. Instead, next-generation sequencing ligated specific sequencing adapters to the ends of each fragment. These sequencing platform-specific oligonucleotide adapters enabled the fragments to be captured, separated, amplified *in vitro*, and sequenced, taking the place of plasmids used in shotgun sequencing. The adapters attached to the fragments are captured onto the surfaces of microscopic beads (Roche and Life Technologies) or flowcells (Illumina) to keep each fragment discrete during the sequencing process. On these surfaces, the fragment is amplified *in vitro*, to enhance the strength of the sequencing signal, be it fluorescence (Roche and Illumina) or pH-based (Life Technologies) (**Figure 11.5**). With the elimination of the cloning steps and a streamlining of the workflow, next-generation sequencing provided a means of obtaining bacterial genome sequence data that was quicker, easier to perform, and was less expensive.

Although next-generation sequencing has been a major advance in the field of genomics, and is ideally suited to bacterial genome sequencing, there are some limitations of these technologies. Sanger sequencing has been able to reliably produce continuous stretches of sequence of 1200 bases or more. Next-generation sequencing produces much shorter stretches of continuous sequence data. Although some technologies have increased the sequence information from a single sequencing by synthesis reaction, the **read length**, by 10 times their initial capacity,

Genome

Fragments

With adapters

Bound to surface

Amplification

Sequencing

Detection

Figure 11.5 Next-generation sequencing principles. Most next-generation sequencing technologies use sequencing by synthesis, as established in the Sanger sequencing method, but with modifications. DNA is fragmented and sequencing technology specific adapter sequences are ligated to the ends of the fragments. The adapters are used to attach the fragments to a solid surface (bead or flowcell), as primers for amplification of the fragments, and to initiate sequencing of the fragments. As nucleotide bases are incorporated into the DNA, the identity of each incorporated base is determined via fluorescence or changes in pH within the reaction vessel.

the longest reads from next-generation sequencing are still a third of what is possible with Sanger sequencing, in the range of 250–400 bases. As a trade-off, the amount of data produced from each genome is far greater in next-generation systems, with the **data depth** reflecting a significant increase in the number of times each base is sequenced, increasing the reliability of the data and also producing the needed increase in data to achieve enough overlaps of the shorter read lengths to assemble the sequences. Increased data depth also means that sequence variations within the sequenced sample can be detected. If a base has changed in just 10% of the bacterial population, this change may not be seen if that base has only been sequenced eight times. However, when it has been sequenced 100 times, the change in that fraction of the population is evident in the sequence data due to the depth of the data available on the identity of each base.

Next-generation sequencing has limitations

Despite the *in vitro* amplification steps eliminating the loss of clones and the resulting sequencing gaps, it is quite common for next-generation sequencing to be unable to resolve the data into whole-bacterial genomes. In part, this is due to the shorter read lengths of next-generation sequencing, which can only be partially overcome by the increased data depth. This is particularly true of those genome sequences that contain repetitive regions or that have regions of sequence similarity scattered in the genome. Common repeated features such as the copies of IS elements and prophages within the genome and the rRNA loci that are often present in more than one copy are also problematic regions for assembly. When sequenced as fragments, it is difficult to work out the location of the fragment on the chromosome, since the same DNA sequence is present in more than one location (**Figure 11.6**). This is akin to having a jigsaw puzzle where some of the pieces are identical in shape and color. Regions of the chromosome that have the same, or very similar, DNA sequences confound computational assembly systems that are looking for overlaps in sequence similarity in order to assemble the sequence data into longer and longer stretches of sequence. The sequences cannot be placed properly in the genome and often are seen as the same region rather than separate copies in different parts of the chromosome. When read lengths are shorter than repeated features in the genome, increasing the number of times that the genome is sequenced cannot overcome the inability to sequence fully across the repeated feature (see Figure 11.6).

Bacterial genome-sequencing projects shift focus due to next-generation sequencing limitations

Rather than one circular bacterial chromosome, next-generation sequencing data is often in several **contigs**, segments of contiguous sequence data that have been assembled together. For instance, if a genome sequence has been assembled into 50 contigs, it is not known in what order these 50 contigs belong in the chromosome because there is insufficient overlapping sequence data to complete the assembly. Next-generation sequencing genome projects therefore often leave the data in contigs, rather than pursuing the added laboratory work to manually sequence more information to complete the assembly. These partially assembled genome sequences make it difficult, if not impossible, to assess chromosomal inversions, gene orientation across the chromosome, and other differences in the chromosome that rely on data being assembled into one contig.

Likewise, duplications within the genome are a concern when attempting to assemble whole-genome data, even when looking at Sanger shotgun sequencing. Assembly relies on finding overlaps of identical stretches of sequence. If a region of the genome has been duplicated then there are two stretches of the genome that are identical to one another, or nearly so. When looking for similar reads, assembly systems will put these together, such that what is actually two copies in the genome looks as though it is only one copy. It should be that this one assembled region has twice as much sequence data as the rest of the genome, since it is in the DNA twice, so it is possible to spot duplications using data depth information, particularly if

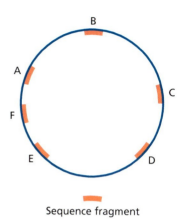

Sequence fragment

Figure 11.6 Repetitive regions complicate assembly. If a genome contains, for example, six repetitive regions that are identical to one another (A, B, C, D, E, and F), this can confound attempts to assemble sequencing fragments. The fragment shown could belong in any of the six regions of the genome.

8. Although genome sequencing is now inexpensive, it seems counterproductive to sequence the same bacterial strain again. What could justify resequencing a genome that has already been sequenced once?

9. What is the difference between genome sequencing and metagenomic sequencing? What might be learned from a metagenome?

10. Two different kinds of single-molecule sequencing have emerged. What is the main advantage of these methods over other sequencing technologies?

11. Of the sequencing methods described in this chapter, which do not use sequencing by synthesis? How do they work instead?

12. What is the difference between a contig and a scaffold? How are each formed?

Discussion topics

These topics are presented for discussion in study groups, as part of class discussions, or on your own. These questions go beyond what is directly covered in this part of the book. Use the research literature and other reading to explore these topics in more depth. Tips to help prepare for topic discussions are available on-line.

13. Describe how next-generation sequencing platforms such as 454, Illumina, and Ion Torrent built upon and developed what had previously been done by Sanger sequencing methods. Discuss common features between these next-generation sequencing technologies and also common features shared with Sanger sequencing. How do these relate to biological systems and processes?

14. Explore the ways in which next-generation sequencing has been used recently in bacterial epidemiology, either on a global scale or locally within a community or hospital. Assess the increased resolution possible using genomic data to demonstrate relationships between isolates and form timelines and theories of transmission events.

15. Resequencing to confirm that mutants constructed contain only the desired mutations, Tn-Seq experiments, ChIP-Seq experiments, and evolve and resequencing experiments have made important contributions to our understanding of bacterial genetics and genomics. Explore the details of one such study and discuss the methods used, how they were chosen to appropriately address the research question, and what they found.

Further reading

The first genome sequence of a free-living organism was bacterial

Fleischmann RD, Adams MD, White O, Clayton RA, Kirkness EF, Kerlavage AR et al. Whole-genome random sequencing and assembly of Haemophilus influenzae Rd. Science. 1995 Jul 28; 269(5223): 496–512.

Loman NJ, Pallen MJ. Twenty years of bacterial genome sequencing. Nat Rev Microbiol. 2015; 13: 787–94.

Early genome sequencing of bacteria provided opportunities for new insight and innovation

Himmelreich R, Plagens H, Hilbert H, Reiner B, Hermann R. Comparative analysis of the genomes of the bacteria Mycoplasma pneumoniae and Mycoplasma genitalium. Nucleic Acids Res. 1997; 25(4): 701–12.

Next-generation sequencing and epidemiology

Harris SR, Feil EJ, Holden MTG, Quail MA, Nickerson EK, Chantratita N et al. Evolution of MRSA during hospital transmission and intercontinental spread. Science. 2010; 327(5964): 469–74.

Bacterial genome sequencing can uncover bacteria never before studied

Quince C, Walker AW, Simpson JT, Loman NJ, Segata N. Shotgun metagenomics, from sampling to analysis. Nat Biotechnol. 2017 Sep 12; 35(9): 833–44.

Quick, easy sequencing means bacterial genomes can be given a second look

Churchward CP, Calder A, Snyder LAS. Mutations in Neisseria gonorrhoeae grown in sub-lethal concentrations of monocaprin do not confer resistance. PLoS ONE. 2018; 13(4): e0195453.

Good BH, McDonald MJ, Barrick JE, Lenski RE, Desai MM. The dynamics of molecular evolution over 60,000 generations. Nature. 2017 Nov 2; 551(7678): 45–50.

Bioinformatics

Bioinformatics is a complex, multidisciplinary field that focuses on the analysis and interpretation of biological data. In the context of bacterial genetics and genomics, bioinformatics refers to the investigation of nucleic acid sequence data and amino acid sequence data. Bioinformatics brings together computer science, mathematics, statistics, engineering, physics, chemistry, and biology. Primarily, bioinformatics is computer-based, however the investigations pursued on the computer can then be brought into the laboratory for verification or further experimentation. In this chapter, the power of bioinformatics in interpreting bacterial data will be explored, including examples of how bioinformatics can be applied to various research questions.

A lot can be learned from looking at strings of A's, T's, G's, and C's

At its most basic, a bacterial genome sequence is a few thousand to a few million adenines, thymines, guanines, and cytosines. Taken as one continuous sequence, these four letters represent the chemical makeup and order of molecules present in the genomic DNA. Hidden in the nucleic acid order are all of the genes that encode the bacterial proteins, all of the sequences involved in regulation of expression of those proteins, and the sequences of all RNA molecules present within the bacterial cell. Using bioinformatics, the locations of the features within the genome can be revealed.

There are no markers that set out definitively where genes occur in the genome based on the DNA sequence A's, T's, G's, and C's alone. To find the genes among the bases, bioinformatics makes a best guess at gene location. These guesses are based on laboratory evidence regarding features such as the nucleic acid sequence of the initiation codons (ATG, GTG, TTG) and the termination codons (TGA, TAG, TAA), however, these sequences can be present in the genome without being initiation or termination codons. In addition, laboratory experiments have revealed the consensus sequences of promoter regions, particularly those recognized by common sigma factors. This can help **bioinformaticians**, those who do bioinformatics, in predicting the locations of promoters and matching them up to predicted locations of genes. These predictions can be further strengthened by identifying the potential ribosome binding sites between the predicted promoter and the predicted CDS (**Figure 12.1**).

The DNA binding sites of some regulatory proteins have been determined in the laboratory, which can also be fed into bioinformatics analysis to predict possible binding site locations. This may reveal which CDSs are involved in the **regulon** controlled by the transcriptional regulatory protein. The various CDSs that are co-regulated by the same regulatory protein can give insight into the functions they conduct together or the conditions under which expression is beneficial. Research into RNAs within the bacterial cell has also enabled bioinformatics predictions of the locations of the ribosomal RNAs, the transfer RNAs, and the non-coding RNAs in the genomic sequence data.

Bioinformatics is essential for interpreting sequence data

Once sequence data is created, the goal is to interpret the data and make it available in public databases in a format that is accessible to other researchers so that the

Figure 12.1 Bioinformatics can predict potential genes and their associated sequences. Using information on initiation and termination codons, it is possible to predict the location of potential genes, as CDSs. Starting with just the DNA sequence data (top), potential initiation and termination codons can be identified. From this a predicted CDS is identified (second from top). Additional support for the predicted CDS comes from identification of a predicted promoter sequence (third from top) and from a predicted ribosome binding site (RBS; bottom).

biological meaning of the sequence data can be understood in the context of the bacterial organism. To achieve this, bioinformatics is essential. The output of sequencing methods must first be processed before the search for genes and RNAs can occur. The **raw sequencing data** are the results of the experimental method for determining the order of nucleotides in a molecule of DNA. All sequencing methods are prone to errors and are less robust in their assessment of the sequence of the DNA as the read length (see Chapter 11) approaches the limits of the technology. There are also instances where a particular sequencing reaction will fail and not produce good quality data. Therefore, a first step for analysis of sequence data is to address the quality of the data and to trim the data down to sequence information that is believed to be correct. If bad data is left in the sequence, it may result in bad predictions later on down the line of investigation, therefore, it is important that the trimming and assessment of data quality is done well (**Figure 12.2**).

There are cases where the sequence of interest is shorter than the length limit of the technology and the sequence can be determined with one sequencing reaction. In these cases, generally an automated Sanger approach is taken, and to increase the robustness of the sequence data achieved, two sequencing reactions will be used, one for each strand to sequence each strand of the DNA. This provides a check of each base determination using two different sequencing reactions that are sequencing the **reverse complement** sequence of one another (**Figure 12.3**), being the sequence on the opposite strand where it is both in reverse and the complementary bases. In these cases, after trimming, a quick comparison of the two sequences and resolution of any differences between them is all that is needed.

For most sequencing efforts, more than one sequencing reaction, or double-stranded sequencing, is needed to achieve the identification of all of the bases of the relevant DNA. This may be because the region of interest for Sanger sequencing is longer than the read length that can be achieved. It may also be that the goal is

Figure 12.2 Raw sequencing data needs to be trimmed. Raw sequencing data contains errors due to the limits of the sequencing technology and the chance that some sequencing reactions do not work well. The quality of the interpretations made later depends on the quality of the data, therefore, poor-quality sequences are removed before the sequencing data is analyzed. The top read (a) has poor-quality sequence at the end, therefore it is trimmed, removing the poor-quality sequence and producing a trimmed read that is included in later analyses. The bottom read (b) is poor quality throughout, therefore, the whole read is deleted and not included in further analyses.

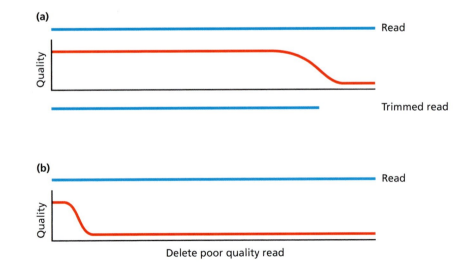

Figure 12.3 Double-strand sequencing for small areas of interest increases accuracy. For a small region of interest that can be sequenced within the length limits of the sequencing methodology, it is possible to collect the needed sequence data with a single reaction. However, sequencing the DNA in both directions (sequencing reaction 1 and sequencing reaction 2) to sequence each of the strands of DNA will produce more robust sequence data (double-stranded data), having used two reactions that run in opposite directions to one another.

the generation of a complete genome sequence. Either way, some **assembly** of the nucleic acid sequence data fragments into longer pieces of sequence is required. This will involve sequencing data that overlaps with other sequencing data from the same sample, such that the overlaps can build upon each other and increase the length of known sequence information (**Figure 12.4**). The resulting assembled sequence information is a contig.

Annotation predicts features in sequence data and notes their locations

An **annotation** is the labeling of features. Annotations are used in books, computer programs, and sequence data. In the context of DNA, the annotation describes the sequence, usually including the bacterial species and strain, as well as information as to the nature of the sequence, be it a short segment, plasmid, or a whole genome. An annotation will include the base numbers of particular features in the sequence. Some features, like CDSs, will be specific to one strand of the DNA, information that is also included in the annotation. The names of identified genes and RNAs and notations as to the attributes of the features in the sequence data will be annotated. These can include the amino acid sequence of the predicted protein product of a CDS, the predictions of subcellular locations of potential proteins, and other key pieces of information of relevance to the features within the sequence. Annotation is not just for genomes. Short segments of DNA and plasmids are also annotated to point out the features in the DNA or those believed to be in the DNA based on analysis (**Figure 12.5**).

A bacterial sequence annotation seeks to provide information on the CDSs and genetic features in the DNA. CDS annotations will include both genes with experimentally defined functions and sequences that look like they may be genes. Annotation of a feature such as a CDS can be as simple as noting its position within a stretch of DNA, but can also provide more information. The predicted protein sequence is often provided for CDSs and is based on the DNA sequence using a codon table to generate the amino acid sequence. If the sequence data suggests that there may be a signal peptide as part of a protein, then this can be noted in the annotation. The predicted amino acid sequence enables prediction of the molecular weight, isoelectric point, cellular location, and domains within the encoded protein. For CDSs that are known genes with experimentally derived functions, then the gene name can and should be included in the annotation. If the function or predicted function is enzymatic, an enzyme classification number can be included in the annotation. This is a numerical identifier for an enzyme based on its activity, identified as being similar in function regardless of feature names or sequence similarity. Annotations should include reference to experimental work, particularly where a feature is a gene with functions described in the literature. Some annotations also include GO (gene ontology) function, a set of terms for describing gene products, and COG (clusters of orthologous groups) designation, a database of similar protein sequences and their assigned functions. If the CDS described is a paralogue or orthologue of other annotated CDSs, this information may be valuable to include. It is possible that the sequence data may include **pseudogenes**, which

Figure 12.4 Assembly of sequence data relies on overlaps in the sequence information from each reaction. Each individual sequencing reaction generates a string of bases that are known to follow one after the other, regardless of whether they are on the forward or reverse strand. However, these may not cover the whole of the sequence desired. Therefore, several overlapping sequencing reactions are needed. These generate data where the overlaps allow assembly of the data into longer stretches of sequence. Once assembled, the contiguous sequences from different sequencing reactions are referred to as contigs.

Sequencing length 329

5–319 coding sequence

5–319 gene *expL* (an example gene)

5–319 amino acid sequence MRGLKNRG···

Figure 12.5 Gene annotation. The annotation for an example DNA sequence of 329 bases, where the coding sequence starts at the fifth base and ends at base 319. This is a known gene, *expL*, with a predicted amino acid sequence (truncated here).

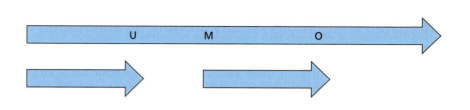

Figure 12.14 Selenocysteine and pyrrolysine codons and sequence annotation. A coding sequence (top line) containing a selenocysteine codon (U) and a pyrrolysine codon (O) could easily be annotated incorrectly. Identification of initiation codons and termination codons to define the coding sequences in the region incorrectly identifies two coding sequences, both much shorter than the original and neither containing the amino acids selenocysteine or pyrrolysine. The first uses the correct initiation codon, but takes the selenocysteine codon as a termination codon. The second uses an internal methionine codon (M) as an initiation codon and takes the pyrrolysine codon as a termination codon. Neither uses the true termination codon, which is after the pyrrolysine codon.

As discussed in Chapter 7, there are genes that make use of the SECIS and PYLIS sequences to change a termination codon into a selenocysteine or pyrrolysine, respectively. Annotations often overlook these due to the nature of the progression of the annotation, which starts with ORF definition. Since the selenocysteine and pyrrolysine codons are stop codons, altered to encode these amino acids via the SECIS and PYLIS sequences, the ORF will be incorrectly defined at the start of the annotation process (**Figure 12.14**).

Naming genes is not straightforward, with some genes having more than one name

During the course of research into the functions of genes, it is very easy for those that are orthologues to end up with different gene names in different species or even different gene names used by different laboratories to refer to genes in the same species. In some annotations, gene names are absent and only functional data for CDSs is included, however, others attempt to place gene names with sequences where possible. This can be problematic when there is more than one name for the gene. In a good annotation, where a gene is known in the literature by more than one name, the synonym will be listed in the annotation.

Complicating annotation of gene names are instances of the same name being used for different genes that have different functions. During the course of research, a research project will name a gene being investigated and use that name in reporting their findings in the literature. It may be that this name is already in use in this species or that the name is used in another species. When the whole of the genome is assessed for homologues and annotated, CDSs are assigned the same name using information from different sources, which can result in genomes having more than one gene with the same gene name.

Annotation errors, including spelling mistakes, can spread from one annotation to many others

Annotations rely on data from other annotations in the databases. Homology searches identify similar sequences in other annotated data and these other annotations contribute to new annotations. The result is that each annotation has the potential to be propagated across many other bacterial sequences. Errors in one annotation can expand to be errors in many, with the incorrect information spreading. This is most obvious in the propagation of spelling errors within annotations. While it may seem trivial to worry about spelling errors when some annotations are simply wrong in their identification of sequences and their function, spelling errors present a problem for analysis. For example, if the function of a CDS is not spelled correctly, this makes simple text searches for genes of that particular function impossible. It is possible for internet search engines to identify potential spelling errors and suggest to the user that they have misspelled something. Incorporation of such programing into the tools used in genetics and genomics would be useful for the identification of annotation spelling errors.

Gene locus identifiers are handy for labeling features in annotations, but reveal nothing about function

The original annotation of *Escherichia coli* strain K-12 assigned gene names to all of the annotated CDSs. Genes of known function with previously assigned names

Figure 12.15 Locus identification numbers indicate relative locations of annotated CDSs. Annotation assigns a locus identification number to each identified coding sequence in numerical order along the sequence. To these locus identification numbers, additional annotation information can be added, such as gene names or identification as encoding a hypothetical protein (hypo). The locus identification numbers allow the sequence to be unambiguously referred to, whereas there may be multiple sequences annotated as *abcB* in the databases.

were annotated as such, but those that were not previously named were given names starting with *y*. This system was used in the early days of bacterial genomics to identify features and relative locations in the genome. These *y* gene names, however, tell us nothing about the functions of the CDSs.

CDS identifiers used in genome annotations are not strictly gene names. Gene names are based on the function of a gene and should not be applied to CDSs. Therefore, the *y* names should not be used in any annotation outside of *E. coli* strain K-12. In such an instance where homology with one of these CDSs is noted, it is acceptable for the annotation to note the homology, but the *y* name should not be used as a gene name in other annotations. In *E. coli*, when a function is identified experimentally, the *y* gene designation should be replaced with a name related to the gene's function.

Later annotations of bacterial genomes used **locus identification numbers**. These are combinations of letters, to specify the bacterial genome, and numbers, to indicate the relative location on the chromosome. Locus identification numbers are simply markers for CDSs and unambiguously refer to a single feature in the genome (**Figure 12.15**). In most annotations, features will have both the locus IDs and gene names, where applicable. As with the *E. coli y* names, the locus identification numbers should not be transferred across to another annotation's gene name. The identifiers from other genome sequence annotations can go into the notes in an annotation to indicate homology.

During the course of annotation, identified CDSs are numbered sequentially along with a specific tag to identify the genome sequence itself (see Figure 12.15). These locus identification numbers are not gene names, but can clearly identify a particular CDS from a particular genome sequence.

There are three major public databases for sequence data

Sequence data, either nucleic acid or amino acid, is available via public databases. Sequences that are published in the research literature are required to be deposited into one of these databases so that they are freely available to other researchers. There are three major public databases for sequence data, which together form the International Nucleotide Sequence Database Collaboration: the National Center for Biotechnology Information (NCBI) GenBank; the European Bioinformatics Institute European Nucleotide Archive (EBI ENA); and the DNA Data Bank of Japan (DDBJ). Each of these centers provides a repository for sequence data and searchable databases for DNA and RNA data, as well as protein sequence data and other information. The data contained at each of the three centers is synchronized daily to ensure that they carry the same information.

While the data in GenBank, ENA, and DDBJ are the same, each database presents the data in its own format. One of the features of the synchronization is to transfer the databank entry from one database's format to the format of the database taking in the data (**Figure 12.16**).

Comparative genomics finds that there are commonly shared genes and unique genes

In comparative genomics, the genomic features of different organisms are analyzed against each other. This can include the similarity between and the presence of the sequences of the CDSs, ncRNAs, and other features. Slight changes in regulatory

Mapping of the sequence reads against a reference genome can reveal small-scale genetics differences, but it is harder to identify larger regions of difference using this technique. Large differences will not align with the reference genome, therefore, to reveal these it is necessary to look at the unaligned sequencing data to see genetic features that are not in the reference genome.

Horizontally transferred gene sequences tend to carry a signature that can identify them

Horizontal gene transfer is a key characteristic of bacteria, enabling the exchange of genetic material between strains and species. Horizontally acquired genes and DNA can also be called foreign genes and foreign DNA. DNA signatures of horizontally transferred sequences may be different from the rest of the genome in features like %G+C and codon usage (Chapter 3), as well as perhaps in amino acid usage, where using an amino acid in a protein sequence is rare for a particular species. Using bioinformatics, DNA signatures of the chromosome are analyzed to identify regions that are potentially horizontally transferred. A few bacterial species are naturally competent for transformation, like *Neisseria meningitidis* and *Neisseria gonorrhoeae*. For these species, there is extensive bioinformatics and experimental evidence of horizontal gene transfer in genome sequence data. Evolution in such species is heavily influenced by the ability to acquire DNA.

Recently acquired sequences are more likely to be different from the rest of the genome (**Figure 12.20**). Over time, sequences **ameliorate**, which means they change to match the genome with regard to %G+C and codon usage. It is therefore easier to identify recently horizontally acquired sequences than it is to identify those that have been within the bacterial genome for some time. There are some regions of genomes that may appear to be foreign when they are not. These regions are native to the genome, but due to the nature of the features in a region, it may look horizontally transferred based on its %G+C and codon usage.

Horizontally acquired regions can also be identified using phylogeny. Identification through phylogeny relies on orthologues being identified between sequences. Those that have no orthologues could be foreign (**Figure 12.21**). Alternatively, the identified genes may be strain-specific because they were generated via duplication and change or deletion rather than horizontal gene acquisition.

With the focus on genomes and genomic sequence analysis, it could be easy to overlook the role that plasmids play as sources of horizontally transferred genes; plasmids can, in whole or in part, have originated from other bacteria. As extra-chromosomal DNA, plasmids are capable of self-replicating and of carrying with them the regulatory features needed to express the genes carried on the plasmid. Therefore, plasmids are efficient transmitters of horizontally transferred genes. Bacteriophages are also effective transmitters of horizontally transmitted genetic material (Chapter 3). In addition to the prophage sequences themselves being horizontally acquired, bacteriophages may bring foreign genes with them into the genome, contributing to horizontal gene transfer between bacteria.

Figure 12.20 %G+C can be an indicator of horizontal gene transfer. Region 1 has a clear change in the %G+C for part of the sequence, which corresponds to some horizontally transferred DNA (HT). In region 2, the horizontally transferred DNA has been present in the genome for some time and has ameliorated to more closely match the %G+C of the rest of the region. Region 3 includes the rRNA locus, which has a %G+C signature that is different from the rest of the genome, by nature of the function of the sequence.

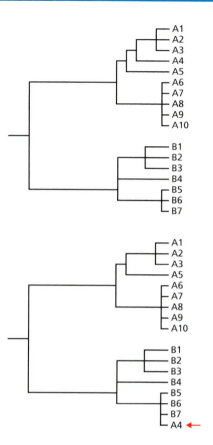

Figure 12.21 Phylogeny can reveal horizontally acquired sequences. This phylogenetic tree (top) shows two species, labeled A and B, with several strains representing each, all numbered. When a single gene is considered (bottom), it is revealed that the gene from A4 clusters with the genes from species B rather than with the other species A genes.

Genome sequence analysis can find unexpected features in the sequence data

Prior to whole-genome sequencing, many assumptions were made about the nature of genetic material within bacterial cells and our understanding of it. Long held beliefs such as bacteria having only circular chromosomes were overturned with the sequencing of linear chromosomes from *Borrelia burgdorferi* and other bacteria, confirming previous evidence. Flagellar gene systems were found within intracellular bacteria such as *Buchnera aphidicola* that do not express flagella yet use some of the flagellar proteins to form structures that transport proteins across the bacterial cell surface. Bacterial species were found to contain horizontally acquired eukaryotic DNA and vice versa. With important implications in our understanding of bacterial pathogens, virulence genes have been found in non-pathogens. These and other genomic analysis-based results have brought to light ecological and evolutionary contexts for gene presence in bacteria. Findings such as these have required a re-examination of the potential phenotypic influence the presence of a gene or set of genes within a genome may have upon the characteristics of the microorganism.

Genome sequence data has enabled the identification of signatures for uptake signal sequences enhancing horizontal gene transfer that have not previously been identified. For example, in the *Apl* subclade of *Pasteurellaceae* including *Actinobacillus pleuropneumoniae*, *Mannheimia haemolytica*, and *Haemophilus ducreyi*, horizontal gene transfer is enhanced between one another when the ACAAGCGGT sequence is present in the DNA. This sequence signature was identified through analysis of the sequence data for short sequences present in several copies around the chromosome at a frequency well above what might be expected by chance. It is different in both sequence and specificity from the AAGTGCGGT uptake signal sequence in the *Hin* subclade, which includes *Haemophilus influenzae*. Likewise, analysis of the genome of *Helicobacter pylori* revealed that there is no uptake signal sequence in this species, the assumption being that the acidic niche of the bacteria acts as a restriction to horizontal acquisition adequately.

As the understanding of bacterial genomes and genomic features has progressed, it has been possible to identify within the genome sequence data features that were never before recognized. Perhaps the most apparent of these is the identification of small non-coding RNAs in regions of the chromosome previously believed to be non-functional junk. In addition, genomic analysis has become better at finding repeat regions that might have regulatory functions, might be hotspots for recombination, or that otherwise might be of interest, such as the CRISPR repeat regions that will be discussed in Chapter 13. With the inclusion of more sequence data, it has become easier to identify prophages and prophage fragments within bacterial genomes, particularly those that share homology with prophages in other genome sequence data, even from unrelated species. Other integrated features such as IS elements, transposons, and fragments of these features in the genome, as well as other regions that are potentially related to these elements, can now be readily identified.

Comparative genomics on closely related strains can reveal biologically important information

A key experiment in bacterial genetics is to knock-out a gene and identify the change in phenotype. This requires laboratory investigation and it may be that the phenotype involved is not readily apparent. It is possible to compare the phenotypes of very

Comparative genomics on closely related strains can reveal biologically important information

Snyder LAS, Loman NJ, Linton JD, Langdon RR, Weinstock GM, Wren BW, Pallen MJ. Simple sequence repeats in *Helicobacter canadensis* and their role in phase variable expression and C-terminal sequence switching. *BMC Genomics* 2010; 11: 67.

Zelewska MA, Pulijala M, Spencer-Smith R, Mahmood HA, Norman B, Churchward CP, Calder A, Snyder LAS. Phase variable DNA repeats in *Neisseria gonorrhoeae* influence transcription, translation, and protein sequence variation. *Microb Genomes.* 2016; 2(8): e000078.

Function predictions can be made based on sequence similarity

Altschul SF, Gish W, Miller W, Myers EW, Lipman DJ. Basic local alignment search tool. *J Mol Biol.* 1990; 215: 403–10.

Part V

Bacterial Response, Adaptation, and Evolution

In order to survive in a changing world, all organisms must themselves be able to change, to respond to the ways in which their environment changes, to adapt to those changes, and to evolve to take best advantage of the resources available to them. Bacteria have been responding, adapting, and evolving for millions of years as our planet has changed and as new niches have developed. Over time, eukaryotic organisms developed, presenting bacteria with host organisms upon which they could be symbionts, pathogens, or **saprophytes**, living on dead organic matter. In this section, the concepts of bacterial response, adaptation, and evolution will be explored in greater depth. These three concepts are very much overlapping, and some aspects will be covered in more than one chapter, such as quorum sensing, which is addressed in this chapter and in Chapter 14. This section will explore how bacteria respond to their environment, contributing to their phenotype and how this contributes to their abilities to adapt. The evolution of these responses and adaptations will then be discussed in the final chapter of the section.

Figure 13.9 Bioluminescence of the bobtail squid by *Aliivibrio fischeri*. The squid has within its body a hospitable niche where the bacteria can grow and divide, reaching a large population size. When the bacterial quorum sensing system detects that there are enough bacteria present, the genes for bioluminescence are switched on, producing the characteristic glow of the bobtail squid. (Courtesy: wildestanimal/Shutterstock.com.)

predicted at the time that research into why the bobtail squid glows at night would lead to insights as to why catheters and other medical devices can become colonized with life-threatening infections. Both are cases where bacteria are responding and expressing certain genes only in cases where their populations are large. Both are very closely related quorum sensing systems, yet the connection between them could not have been predicted when the research began.

Quorum sensing makes a beautiful bioluminescent glow in the ocean and in the lab

Some sea creatures, such as the bobtail squid produce a beautiful bioluminescent glow. These squid have special organs called **photophores** or light organs in their bodies that become colonized by bioluminescent bacteria such as *Aliivibrio fischeri* (formerly *Vibrio fisheri*). The squid have evolved to have this organ as a hospitable environment for the bacteria so that the sea creature can then harness the light production of the bacteria, which prevents the squid from casting a shadow at night that might give away its presence to predators. These two organisms therefore live in **symbiosis**, a long-term interaction between two different species. The *A. fischeri* bacteria do not produce their bioluminescent glow all of the time. Making light is metabolically expensive and there is little to no benefit of doing so when there are only a few bacterial cells, since there will be little to no light produced. Indeed, during the day when the light is not needed, the squid purges its light organs of most of the bacteria, thereby turning off bioluminescence. When quorum is reached, when there is a large population of bacteria present in the photophore organ, then the *A. fischeri* bacteria glow (**Figure 13.9**).

Quorum sensing is a process to tally the bacterial cells

Quorum sensing allows bacteria to tally the number of bacterial cells in the vicinity and discover how many other bacterial cells of the same species there are, assessing the presence and concentration of bacterial cells. This process has, essentially, three main parts. In the first, the bacteria make and secrete an **autoinducer**, which is a molecule that the bacteria can recognize. The types of autoinducers used vary by species and some bacterial species produce more than one type of autoinducer. The second part is to detect the changes in the external concentration of autoinducer. As more bacteria are present, there should be more autoinducer detected in the environment. In some species, autoinducers from other species can be detected, even when they themselves do not make and secrete autoinducers. In the third and final part, the bacteria regulate their gene expression based on the detection of the autoinducer. The regulatory proteins and genes regulated in response to quorum sensing differ by species and the requirements for expression of genes as the bacterial population grows. Once the amount of autoinducer reaches a certain level, this indicates that the bacterial population has reached the threshold for gene expression regulation (**Figure 13.10**).

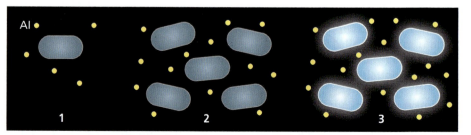

Figure 13.10 Quorum sensing has three parts. (1) The cells make autoinducer (AI). (2) The cells sense autoinducer. (3) The cells respond to a threshold amount of autoinducer that indicates there are enough other bacterial cells around to turn on advantageous genes, such as those for bioluminescence.

Autoinducers are abbreviated as AIs and most in Gram-negative bacteria tend to be molecules like acyl-homoserine lactones, synthesized by LuxI (**Figure 13.11**). These types of AI-1 molecules are often able to diffuse freely across the bacterial membrane, allowing them to leave the cell, accumulate in the environment, and diffuse into other bacterial cells. Some larger variants of AI are actively exported from cells by efflux pumps. There are specific receptors for Gram-negative AIs that are either in the inner membrane or cytoplasm, which recognize the AI structure. When the Gram-negative AIs bind to these receptors, they are able to trigger regulation of gene expression of the Gram-negative quorum sensing regulon. This regulon is the set of quorum sensing regulated genes that are orchestrated together as one system, with some genes being activated and some being repressed. In addition, the quorum sensing regulon often includes the genes encoding proteins for biosynthesis of the signaling molecule.

In Gram-positive species, the quorum sensing autoinducers are peptides of various lengths and modifications, autoinducing peptides abbreviated as AIPs (**Figure 13.12**). When there is high cell density of Gram-positive bacteria, the AIP binds to a two-component histidine kinase, which autophosphorylates and then phosphorylates and activates the response regulator. This response regulator activates transcription of the quorum sensing regulon. Some Gram-positive AIPs can enter back into cells as well as being present in the environment and in this way the AIP can directly interact with the regulation of gene expression, as well as through the two-component system.

Quorum sensing can involve not just monitoring the growth of the bacteria's own population, but also assessment of how many other bacteria there are around of other species. Interspecies quorum sensing involves a different autoinducer, AI-2 made by LuxS. Although there can be some chemical isomers and variations on the base structure, generally AI-2 is *S*-2-methyl-2,3,3,4-tetrahydroxytetrahydrofuranborate (**Figure 13.13**). The cross-species communication possible through AI-2-mediated quorum sensing enables multi-species biofilms to develop and mature, however other gene systems can be controlled by AI-2 signals as well. Indeed, AI-2 was first discovered through research into bioluminescence in *Vibrio harveyi*. It was observed that supernatant from cultures of other bacterial species were able to induce bioluminescence in *V. harveyi*.

Biofilms mature due to quorum sensing signals

Biofilms are dense bacterial populations that form on surfaces in a matrix that the bacteria generate of extracellular polysaccharides, DNA, and proteins. Perhaps, one of the most important realizations to come from the discovery of quorum sensing was that the development of biofilms is orchestrated by the signals sent between bacteria via autoinducers within the growing population. The maturation and development of the bacterial biofilm relies heavily on quorum sensing signals and its disruption can be influenced by disrupting quorum sensing. In several species, experimental studies have demonstrated that without functional quorum sensing systems, bacterial populations cannot form effective biofilms. Due to their placement within the biofilm, some cells will experience limited oxygen and nutrients, therefore, careful regulation of genes is necessary for the survival of the population.

One of the issues with biofilms is that they are notoriously difficult to eradicate, being **tolerant** to antibiotics. Bacteria have evolved biofilms to enable their survival and this also happens to confer **tolerance** to the antibiotics that humans have started using against them. This means that although there is no genetic basis for resistance, there are physiological properties of the biofilm that are protective of the population as a whole that prevent killing of all of the bacteria by the antibiotic. Due to their properties, biofilms just happen to confer antibiotic tolerance. Regardless, it is believed that the density of the biofilm itself contributes to the observed tolerance of biofilms to antibiotics, as does perhaps the complexity of the extracellular matrix. However, antibiotics can be seen to penetrate through the biofilm, so this cannot be the whole of the reason that biofilms are tolerant to antibiotics. It is likely that the density of the biofilm and the extracellular matrix provide protection against the host immune

Figure 13.11 Acyl-homoserine lactone from *Aliivibrio fischeri*. Gram-negative AI-1 autoinducer molecules can vary in their structure, particularly the length of the side chain and substitution on the acyl chain.

Figure 13.12 Gram-positive auto-inducer molecules are usually short peptides. Unlike Gram-negative bacteria, which make organic molecules based on ASLs, the AIP autoinducers used by Gram-positive bacteria are short peptide molecules. They can be of various sizes and amino acid sequences, which may be further modified, and are recognized by two-component histidine kinases on the surface of the bacteria.

Figure 13.13 Autoinducer-2, AI-2, the molecules of communication between bacterial species. AI-2 is made by LuxS. Different bacterial species can receive AI-2 from one another, thereby communicating as one microbiome community.

7. Give an example of a gene or set of genes that it is important for a bacterial species to express only when a bacterial population has reached a critical threshold size. Explain why this gene or genes should not be expressed in a smaller population.

8. How many different types of autoinducers, in general, have been described? Which types of bacteria, for the most part, make each of these autoinducers and sense them?

9. Describe the major stages of the development of a biofilm and the key processes involved in each stage.

10. What is the difference between antibiotic resistance and the antibiotic tolerance exhibited by persister cells in biofilms?

11. When a bacterial cell is infected by a bacteriophage, what system protects the cell and how does it do this?

12. What can we learn about the history of the bacterial cell by the spacer sequences between the repeats in the CRISPR array?

Discussion topics

These topics are presented for discussion in study groups, as part of class discussions, or on your own. These questions go beyond what is directly covered in this part of the book. Use the research literature and other reading to explore these topics in more depth. Tips to help prepare for topic discussions are available on-line.

13. Two-component regulatory systems are often thought of in terms of regulation of gene expression via a DNA-binding protein that is phosphorylated and then binds to promoter regions in DNA. However, response regulators can have other functions as well. Explore and discuss the role of two-component regulatory systems in one bacterial species, where the response regulators are involved in cellular functions other than transcriptional regulation.

14. List some bacterial responses that are regulated by quorum sensing. Associate each with a bacterial species that uses quorum sensing to regulate this response, the quorum sensing genes involved, and the genes regulated. Indicate why this is beneficial to the bacterial population for that species.

15. Regulation of biofilm dispersal is important for bacterial cells, but also is the subject of research into the control of infectious diseases. Using examples of current research as a guide, explore and discuss why it is believed that understanding and perhaps modulating biofilm dispersal signals may be important in controlling biofilm infections.

Further reading

Two-component regulatory systems enable bacteria to respond to their environment

Francis VI, Stevenson EC, Porter SL. Two-component systems required for virulence in *Pseudomonas aeruginosa*. *FEMS Microbiol Lett.* 2017 June; 364(11)

Bacteria decrease host glucose levels to impair the host immune response

Freyberg Z, Harvill ET. Pathogen manipulation of host metabolism: A common strategy for immune evasion. *PLoS Pathog.* 2017 Dec 7; 13(12): e1006669. doi: 10.1371/journal.ppat.1006669.

Some bacteria cheat and let others do all the work with their Type 3 Secretion Systems

Czechowska K, McKeithen-Mead S, Al Moussawi K, Kazmierczak BI. Cheating by type 3 secretion system-negative *Pseudomonas aeruginosa* during pulmonary infection. *PNAS* 2014; 111(21): 7801–6.

Price PA, Jin J, Goldman WE. Pulmonary infection by *Yersinia pestis* rapidly establishes a permissive environment of microbial proliferation. *PNAS* 2012; 109(8): 3083–8.

A response might only be appropriate when the population is large

Rutherford ST, Bassler Bonnie L. Bacterial quorum sensing: Its role in virulence and possibilities for its control. *Cold Spring Harb Perspect Med.* 2012; 2(11): a012427.

Quorum sensing makes a beautiful bioluminescent glow in the ocean and in the lab

Verma SC, Miyashiro T. Quorum sensing in the squid-vibrio symbiosis. *Int J Mol Sci.* 2013 Aug; 14(8): 16386–401.

Quorum sensing is a process to tally the bacterial cells

Thoendel M, Kavanaugh JS, Flack CE, Horswill AR. Peptide signaling in the *Staphylococci. Chem Rev.* 2011 Jan 12; 111(1): 117–51.

Biofilm dispersal is regulated by different elements between different bacterial species

Davies DG, Marques CN. A fatty acid messenger is responsible for inducing dispersion in microbial biofilms. *J Bacteriol.* 2009; 191: 1393–403.

Fedtke I, Gotz F, Peschel A. Bacterial evasion of innate host defences – the *Staphylococcus aureus* lesson. *Int J Med Microbiol.* 2004; 294: 189–94.

c-di-GMP plays a key role in biofilm regulation in *P. aeruginosa*

Chua SL, Liu Y, Li Y, Ting HJ, Kohli GS, Cai Z *et al.* Reduced intracellular c-di-GMP content increases expression of quorum sensing-regulated genes in *Pseudomonas aeruginosa. Front Cell Infect Microbiol.* 2017; 7: 451.

Bacteria have their own immune system to protect them from bacteriophages

Makarova KS, Haft DH, Barrangou R, Brouns SJ, Carpentier E, Horvath P *et al.* Evolution and classification of the CRISPR-Cas systems. *Nat Rev Microbiol.* 2011; 9(6): 467–77.

Bacteria can also adapt to the presence of other bacteria in their niche by trying to kill them. This reduces competition for nutrients and releases the nutrients within the dead bacterial cells. Bacteria, including lactic acid bacteria such as *Lactobacillus*, produce **bacteriocins**. These are antimicrobial peptides of 30–60 amino acids in length and are encoded by the bacteriocin operon. Immunity to the bacteriocin is conferred by an adjacent gene on the operon, protecting the bacteria that make the bacteriocin from their own antimicrobial peptides. Bacteriocins act on the cytoplasmic membrane, generally of Gram-negative bacteria, disrupting the proton motive force essential for bacterial cellular processes, including transport of nutrients across the membrane. It is possible for bacteria to evolve resistance to bacteriocins, generally through efflux or degradation of the antimicrobial peptide, however bacteriocins are a highly effective means for bacterial cells to reduce or eliminate competitor bacterial cells.

GlcNAc has a role as a signaling molecule as well as being part of the bacterial cell wall

The bacterial cell wall, peptidoglycan, is made up of the sugars *N*-acetylglucosamine (GlcNAc) and *N*-acetylmuramic acid (MurNAc). During bacterial cell growth and division GlcNAc is released from the cell due to the necessary remodeling of peptidoglycan that has to happen in order for the cell to expand and for the process of division to occur. The release of GlcNAc into the environment is therefore a signal that bacterial cells are dividing and can trigger neighboring bacterial cells to adapt to the presence of growing and dividing cells in their proximity. Quorum sensing was previously discussed in Chapters 5 and 13, where a signaling molecule was made, released, and sensed as an indicator of population size. In the case of GlcNAc, this is a molecule that is released as part of the process of bacterial cell growth and division. GlcNAc is used as a signaling molecule for bacteria to adapt to signals of growth from each other and other organisms.

For example, the presence of GlcNAc stimulates antibiotic production in soil bacteria. Although we use antibiotics today to combat infectious diseases, these antimicrobial agents originally evolved as mechanisms to enable bacterial cells to kill neighboring bacterial cells that were their competitors. The signal that there are other bacteria around, the presence of released GlcNAc, therefore turns on production of antibiotics. Likewise, the GlcNAc released from growing Gram-positive bacteria induces production of toxins and virulence factors by *Pseudomonas aeruginosa*. Some bacteria take the GlcNAc signal as a sign to leave the niche and find a new place to colonize; *Escherichia coli* decreases the production of its type 1 fimbrial adhesins thereby increasing transmission to new hosts in response to the presence of GlcNAc. The presence of GlcNAc also reduces production of Curli fibers and fimbrial adhesins, important for adhesion and biofilm formation, enhancing dissemination of the bacteria away from the competitors.

Competitor bacteria can be killed with specialized Type 6 Secretion Systems

As seen in Chapter 9, some bacteria use Type 6 Secretion Systems to target competitor bacterial cells; killing them reduces competition for nutrients and releases nutrients from the dead bacterial cells. A Type 6 Secretion System is a multiprotein surface structure made by some Gram-negative bacterial species that is able to directly inject toxins into host cells and bacterial competitors. In some ways this is similar to Type 3 Secretion Systems that inject effector proteins into host cells, however the Type 6 Secretion System can also be used against other bacterial cells and competitor fungal cells, as well as against host cells.

Type 6 Secretion Systems inject toxins into competitor bacteria that degrade bacterial cell walls, disrupt bacterial cell membranes phospholipids, form pores in bacterial membranes, and degrade nucleic acids. As well as the effector proteins that are secreted by the system, bacterial cells with Type 6 Secretion Systems produce an immunity protein that protect the bacteria from its own toxins. Most Type 6

Secretion System effectors target Gram-negative competitor bacteria, which can include related bacterial species.

For example, many of the bacterial species in the Order *Bacteroidales* colonize the human gut at the same time and compete for space and resources. By diversifying their metabolisms they can make use of different nutrients within the niche. However, *Bacteroidales* can also use Type 6 Secretion Systems and secrete toxins to eliminate the competition.

Caulobacter **differentiate between motile and sessile cells**

Caulobacter are a genus of bacterial cells that have a distinct cell cycle, which generates both motile **swarmer cells** that have flagella and stalked **sessile cells** that adhere to surfaces and are non-motile. Cells differentiate from being motile cells to sessile cells, with each cell type providing an adaptive advantage to the bacterial population in the aquatic environment in which *Caulobacter* are generally found. When a sessile cell divides, a sessile daughter cell remains attached to the surface and a motile daughter cell is released to the environment (**Figure 14.2**). The stalked sessile cells are optimized for adhesion to surfaces, nutrient uptake, DNA replication, and cell division, while the motile swarmer cells are optimized for motility via flagella and are able to disperse the bacteria, enabling them to colonize new niches. The swarmer cell can either grow the population in the local area when nutrients are abundant or go to new areas when nutrients become depleted. The swarmer cells also have adhesive pili associated with the flagella; as the swarmer cell transitions to becoming a sessile cell in its site of colonization, the flagella is lost and these pili attach to a surface and retract, bringing the cell close to the surface so that the stalk can form (see Figure 14.2).

The cell cycle of *Caulobacter* is controlled by coordinated expression of the regulatory proteins CtrA, DnaA, GcrA, and CcrM. These are master regulator proteins that control regulons involved in the differentiation of the *Caulobacter* cells. The overlapping of the expression of the different master regulator proteins determines the expression of flagella, the adhesion of the sessile stalked cell to the surface, and the process of DNA replication and cell division (see Figure 14.2). Division generates one sessile stalked daughter cell and one swarmer cell. The stalked daughter cell remains sessile and it is here that CcrM, a DNA methyltransferase acts. Swarmer cells are motile initially, but then become sessile cells as part of the cell cycle, due to the influence of the master regulator proteins (see Figure 14.2).

The control of the genes that regulate the expression and degradation of DnaA, GcrA, CtrA, and CcrM is regulated in an interconnected manner (**Figure 14.3**). In general, proteolysis ensures that the activity of proteins will stop at the right time, through degradation of proteins and therefore their inactivation. Proteins

Figure 14.2 The cell cycle of *Caulobacter* is controlled by master regulatory proteins. The motile swarmer cell expresses the master regulator CtrA and starts to express DnaA. The cell loses its flagella, as shown in the next cell to the right. As DnaA is joined by GcrA, the motile cell becomes a stalked sessile cell and DNA replication begins. Levels of GcrA diminish as CtrA returns and cell division begins, with a flagellated daughter cell forming. The stalked cell remains, able to divide again to produce more flagellated daughter cells. The motile daughter cell is able to move to new areas before becoming sessile and starting a new cycle.

Figure 15.1 Experimental evolution experiments. Richard Lenski's research group has been growing *E. coli* cultures continuously since February 1988. These *E. coli* started from a single culture and have been through over 50,000 generations. The experimental design involves the growth of 12 cultures that are continuously maintained. There are six cultures of *E. coli* REL607, which is Ara$^+$, and six cultures of the isogenic strain *E. coli* REL606, which is Ara$^-$. Each day 0.1 mL of media containing growth from the previous day's culture is transferred into 9.9 mL of DM media with glucose and grown at 37°C with shaking. Each 75th day, freezer stocks are made of each culture and the cultures are checked for contamination by inoculating some of the culture onto TA plates and incubating these at 37°C to confirm that only *E. coli* are growing.

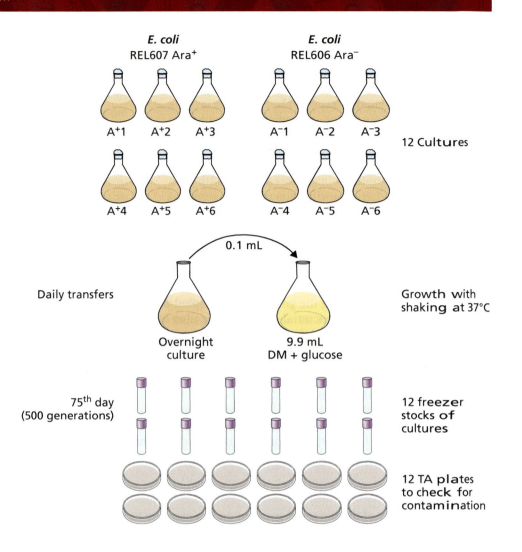

Much of what we know about the rate at which evolution occurs comes from laboratory evolution experiments and from observations of evolution happening in strains isolated from their natural environments. There is no one universal rate at which mutations will occur within a bacterial genome and therefore predictions of bacterial evolution need to be made on a species by species basis and in some cases a strain by strain basis. Variations in mutation rates within a species and the ability to control mutation rates can influence the evolution of different bacterial strains within a species.

Bacteria can evolve within the host and we can see this happen with sequencing technologies

Advances in genome sequencing have revealed two important and previously often overlooked aspects of microbiology. The first is that bacteria can evolve within the host. Several studies have been done that have shown the within-host changes to various bacterial genome sequences over time, revealing the mutations that occur and the phenotypic implications (**Figure 15.2**). In one study, the genome sequences of *Burkholderia pseudomallei* chronic infections from seven cystic fibrosis patients were analyzed over a period of time, with samples collected at about 4 months and 55 months apart. This revealed changes to the genomes including mutations resulting in antibiotic resistance, adaptations to survive in the host, and reduction in virulence leading to persistence of the infection.

The second important aspect of microbial infections previously often overlooked is that a host can be infected with two (or more) different isolates of the

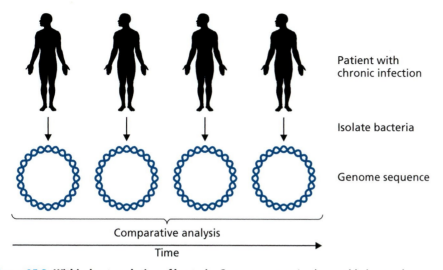

Figure 15.2 Within-host evolution of bacteria. Genome sequencing has enabled researchers to monitor the changes to the bacterial genome that occur within the host during the course of the infection. In patients with a chronic infection (or other long-term infection), the bacteria are isolated at various times during the course of the infection in a single patient and genome sequenced. Whether as complete circular genomes (shown here) or unassembled sequence reads, the data is comparatively analyzed to investigate the within-host evolution of the bacteria.

same species. Often in the diagnostic laboratory, a single organism is identified as infecting the host, yet it is known that patients can be infected with more than one type of bacteria of the same species at the same time. This has become more apparent with whole genome sequencing (**Figure 15.3**). When we look at an infection in detail, often with genome sequencing, we can see that although the patient has the symptoms of one infection, it may actually be caused by more than one strain of the same species.

Together, these two important aspects of bacterial infection set up a tremendous opportunity for within-host evolution. The basic mutation rate of the bacterial species will provide a base rate at which the bacterial species can change within the host. The presence of another closely related bacterium of the same species presents the opportunity for horizontal gene transfer, particularly as they are occupying the same or closely associated niches. If these are bacterial species that are naturally competent for transformation, the opportunity for evolutionary change is even greater. Of course, bacteria also tend to be part of complex microbial communities from which there is the potential to acquire new genetic material. By capturing the genome sequence data at the start of an infection and then later on in a persisting infection, it is possible to track the evolutionary changes to the genome and to identify how the infecting bacteria has diversified within the host, perhaps forming into subpopulations that have and have not mutated certain genes or that have and have not acquired certain genes.

Figure 15.3 Infections can be caused by more than one strain of the same species. When the infection-causing bacteria are isolated from a patient, laboratory diagnostics can identify the species of the bacteria, however this will not identify the presence of more than one strain of the same species in the patient. Genome sequencing of the isolated bacteria can result in assembly of the data into two or more distinct genomes, making it clear that there are multiple infecting strains of the bacterial species.

Antibiotic resistance is an easily observable evolutionary event

Most evolutionary events are slight changes for the bacterial cell that give it a slight advantage over other cells, thus selective pressures favor the new phenotype. Sometimes, evolution makes some striking changes, but most commonly, the changes are very small. Antibiotic resistance is a difference with a big impact that can evolve quite suddenly in a bacterial **lineage**, due to the strong selective pressures of the antibiotics that kill the bacteria without the resistance trait.

Unlike horizontal transfer of an already evolved resistance gene, evolution of antibiotic resistance generally occurs in the opportunistic pathogen and in the non-pathogen part of the microbiome. These are bacterial species that are inhabiting the host organism and otherwise not causing any harm through their colonization at the time of antibiotics exposure. The introduction of an antibiotic to a host organism exposes all of the bacteria within the body to selective pressures, regardless of the intended target of the antibiotic.

Exposure of the microbiome to antibiotics sets up a situation where resistance can evolve within any of the bacterial species within the host organism. Some microbiome species will be naturally resistant to the antibiotic. This could be because they do not have the target of the antibiotic or because the antibiotic is unable to reach the target within the bacterial cell. If the microbiome species is otherwise susceptible, some cells within the population may survive killing by the antibiotic due to selection of pre-existing mutations within the population. Antibiotic resistance can then evolve in these microbiome bacteria (**Figure 15.4**). Although they were not the intended targets to be killed by the antibiotic, because they were in the body treated by the antibiotic, these bacteria end up evolving resistance.

Mutations can be introduced into bacterial DNA by a variety of factors

The key factor to remember about evolution is that the situation does not cause mutations to happen. Mutations are there and are selected for by the situation. The populations of bacteria are exposed to a situation where some of the bacterial cells have a fitness advantage due to a mutation that is already present in those cells (**Figure 15.5**). The situation does not induce the mutation. Mutations are already present in the population. Perfect replication the way it was illustrated earlier in this book (see Chapter 1) almost never happens. Strand breaks occur during replication and have to be repaired; this is in addition to polymerase errors in

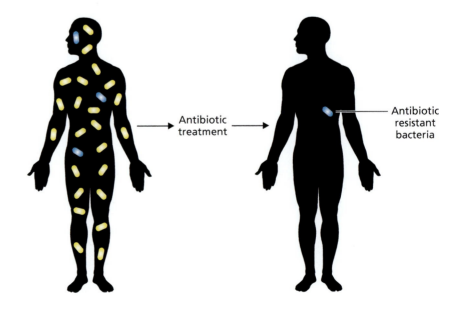

Figure 15.4 Antibiotic resistance evolution in microbiome bacteria. When antibiotics are used against an infection (yellow bacteria), there is a chance that resistance can arise in the bacteria that are part of the microbiome in the body (blue bacteria). Thus, the evolution of resistance can be observed.

Antibiotic treatment

Antibiotic resistant bacteria

replication. Any time a repair has to happen, there is a chance of mutation, despite the efficiency of repair mechanisms.

However, there are a variety of factors that can introduce mutations and some of these factors include situations where it can be beneficial for the bacteria to acquire a mutation in order to gain a selective advantage evolutionarily. For example, DNA can be damaged by UV light and by chemicals. It is essential that bacteria can repair such DNA damage, so that detrimental mutations do not accumulate. In some cases, these mutations can remain uncorrected in the genome and may provide a selective advantage in protecting the bacteria from UV or the chemicals to which the bacteria were exposed. Stress turns on stress responses including molecular mechanisms that cause mutations, which enables **evolvability**, the capacity to evolve adaptations. The general stress response involving RpoS (σ^S) SOS DNA-damage response, membrane stress response, and the stringent response, can all result in mutations. These stress responses regulate mutagenesis and thus regulate the ability of the bacteria to evolve. It must be remembered though, that the exposure did not create the mutation in specific locations to deal with the problem; evolution is a random process of mutation and selection. The mutations happen randomly in the DNA, it is selection that determines if those mutations have a benefit.

Sublethal exposure to some antimicrobials can damage DNA and cause genomic changes. For example, the antibiotic bleomycin binds to DNA and induces double-strand breaks, which may result in mutations. The quinolone class of antibiotics inhibits the DNA replication enzyme topoisomerase ligase domain without affecting the nuclease domain. It therefore cuts DNA without re-ligation, potentially introducing mutations. Production of reactive oxygen species, ROS, in response to antibiotics like beta-lactams, quinolones, and aminoglycosides can also damage DNA. ROS can cause DNA strand breaks and incorporation of oxidized guanine into the genome. Antimicrobials can also trigger the movement of mobile genetic elements present in the genome, which can cause mutations through their introduction into new regions of the chromosome. For example, in *Staphylococcus aureus* sub-inhibitory exposure to quinolones up-regulates the SOS response, which increases IS256 transposition and movement of integrating conjugative elements.

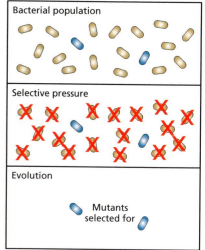

Figure 15.5 Evolution selects for mutations already present in the population, it does not cause mutations. Evolution does not cause mutations to occur in response to selective pressures. Rather, selection acts on the bacterial population and those cells with mutations that give them an advantage (blue) will out compete the other cells in the population.

Yersinia pestis, causing plague, has evolved from Yersinia pseudotuberculosis

Historically, there is great interest in the major infectious diseases that have had a significant impact upon the human population, including the Black Death of the fourteenth century. The bacterial species responsible is *Yersinia pestis*, which has actually caused three major pandemics in recorded history: the Justinian plague (541–542 AD); the Black Death (1346–1353 AD); and Modern Plague beginning in 1894. *Y. pestis* has been investigated at the genomic level to understand the loss and acquisition of genomic content from this species. These data suggest that the acquisition of a number of genomic islands and plasmids led to the evolution of *Y. pestis*, and therefore the emergence of plague, from another species, *Yersinia pseudotuberculosis*.

Since its emergence as a species, *Y. pestis* has lost DNA regions as well as acquired them. These changes have generated different lineages of *Y. pestis*, sometimes referred to as biovars and genomovars. Genomic analysis has determined that modern plague originated in the Yunnan Province of China as *Y. pestis* biovar orientalis. This modern lineage evolved from the *Y. pestis* lineage that caused the Black Death, as determined through genome sequence data obtained from DNA fragments extracted from victims buried in plague pits. Genome sequencing has also confirmed that *Y. pestis* was responsible for the Justinian plague, but that neither the Black Death plague nor modern plague had evolved from this earlier *Y. pestis* lineage.

To obtain the ancient DNA needed in order to investigate the genomics of ancient diseases, a source of DNA that is likely to have resisted degradation is needed. Researchers in this field attempt to obtain samples by drilling into the teeth of skeletons to extract DNA, in the hope that enough intact DNA fragments will remain that can be interpreted once they are sequenced. Since there is very little DNA of the capsule locus by horizontal gene transfer in *Neisseria meningitidis* is often accompanied by the loss of UDP-GalNAc synthesis. *Sci Rep.* 2017; 7: 44442.

Horizontal gene transfer can bring new genes into a species, contributing to its evolution

Wiedenbeck J, Cohan FM. Origins of bacterial diversity through horizontal genetic transfer and adaptation to new ecological niches. *FEMS Microbiol Rev.* 2011; 35(5): 957–76.

The particular nature of an environmental niche can create opportunities for evolution

Bart MJ, Harris SR, Advani A, Arakawa Y, Bottero D, Bouchez V *et al*. Global population structure and evolution of *Bordetella pertussis* and their relationship with vaccination. *mBio*. 2014; 5(2): e01074–14.

Part VI

Gene Analysis, Genome Analysis, and Laboratory Techniques

The next three chapters look at the tools and techniques involved in doing the research that has been discussed in the previous five parts of this book. Here, the discussions will move away from theory. These chapters will delve into the application of skills to the various forms of analysis and experiments that are conducted on DNA, RNA, and proteins. Whether these are the sequence data being investigated on a computer or the physical molecules being investigated at the laboratory bench, the methods involved are included here. First, gene analysis techniques will be explored and then the specialized analysis tools used when scaling up to investigate whole bacterial genomes. Links to web resources referred to in these chapters are available in the on-line supporting materials for this book. Finally, the last chapter in this section is dedicated to laboratory techniques that are key to investigating bacterial genetics and genomics.

and the amino acids of the protein of interest are mapped onto the experimentally determined structure to produce a predicted structure.

Protein threading can suggest a protein structure based on protein fold similarity

Protein threading is useful when there are no homologous proteins with an experimentally determined structure. In this case, there may be proteins that, although not homologous, overall share the same type of **protein fold**. This means that some of the internal structures of the protein are similar to other known protein structures in the way that they fold. This method is also called **fold recognition**. There are only about 1,300 unique folds that proteins are able to make in nature, therefore it is computationally possible to determine potential folds based on amino acid properties and known potential fold patterns. Most new protein structures determined experimentally are based on known folds.

Some gene tools are used to help design laboratory experiments

Primer3 is a program that is used to design PCR primers, sequencing primers, and hybridization probes. The parameters of design can be altered by the user to tailor the resulting primers, or probe, to the particular needs of the laboratory application (**Figure 16.22**). Poorly designed primers are often at the heart of failed PCR

OLIGO	start	len	tm	gc%	any_th	3'_th	hairpin	seq
LEFT PRIMER	35	20	59.11	55.00	25.81	0.00	0.00	cagttggcaacctggattcc
RIGHT PRIMER	510	20	58.94	55.00	19.94	4.76	0.00	ctcacgtacgtctttgctgg

SEQUENCE SIZE: 855

INCLUDED REGION SIZE: 855

PRODUCT SIZE: 476, PAIR ANY_TH COMPL: 0.00, PAIR 3'_TH COMPL: 0.00

```
  1 atgactgacaaaatgcaaagtttagctttagccccagttggcaacctggattcctacatc
                                        >>>>>>>>>>>>>>>>>>>>

 61 cgggcagctaacgcgtggccgatgttgtcggctgacgaggagcgggcgctggctgaaaag

121 ctgcattaccatggcgatctggaagcagctaaaacgctgatcctgtctcacctgcggttt

181 gttgttcatattgctcgtaattatgcgggctatggcctgccacaggcggatttgattcag

241 gaaggtaacatcggcctgatgaaagcagtgcgccgttttaacccggaagtgggtgtgcgc

301 ctggtctccttcgccgtccactggatcaaagcagagatccacgaatacgttctgcgtaac

361 tggcgtatcgtcaaagttgcgaccaccaaagcgcagcgcaaactgttcttcaacctgcgt

421 aaaaccaagcagcgtctgggctggtttaaccaggatgaagtcgaaatggtggcccgtgaa

481 ctgggcgtaaccagcaaagacgtacgtgagatggaatcacgtatggcggcacaggacatg
                <<<<<<<<<<<<<<<<<<<<

541 acctttgacctgtcttccgacgacgattccgacagccaaccgatggcaccggtgctctat

601 ctgcaggataaatcatctaactttgccgacggcattgaagatgataactgggaagagcag

661 gcagcaaaccgtctgaccgacgcgatgcagggtctggacgaacgcagccaggacatcatc

721 cgtgcgcgctggctggacgaagacaacaagtccacgttgcaggaactggctgaccgttac

781 ggtgtttccgctgagcgtgtgcgccagctggaaaagaacgcgatgaaaaaactgcgcgct

841 gctatagaagcgtaa

KEYS (in order of precedence):
>>>>>> left primer
<<<<<< right primer
```

Figure 16.22 Primer3 results. This Primer3 output shows the selection of a pair of PCR primers for the *E. coli rpoH* sequence. Default primer Tm parameters of a minimum 57°C and maximum 62°C were used, producing here primers with Tm's of approximately 59°C. The size of the product was selected as 400–500 bases, producing here a product with this primer pair of 476 bases. Both the left primer (>>>>>) and right primer (<<<<<) are 20 bases in length and their positions in the sequence of *rpoH* are indicated.

Primer pair 1	Sequence (5'->3')	Template strand	Length	Start	Stop	Tm	GC%	Self complementarity	Self 3' complementarity
Forward primer	TTAACCCGGAAGTGGGTGTG	Plus	20	278	297	59.89	55.00	4.00	0.00
Reverse primer	TCCTGCAACGTGGACTTGTT	Minus	20	764	745	60.11	50.00	4.00	2.00

Product length 487

Figure 16.23 Primer-BLAST results. This Primer-BLAST output shows the selection of a pair of PCR primers for the *E. coli rpoH* sequence. Default primer Tm parameters of a minimum 57°C and maximum 62°C were used, producing here primers with Tm's of approximately 60°C. The size of the product was selected as 400–500 bases, producing here a product with this primer pair of 487 bases. Both the forward primer and reverse primer are 20 bases in length. This pair of primers was BLAST searched against the non-redundant nucleotide database entries for *E. coli* to confirm that the primers will specifically anneal to only these locations in the genome; primers that anneal elsewhere are not displayed. The output of this search is displayed to the user, who is able to select sequences that are the intended target for PCR primer annealing.

experiments, therefore the use of Primer3, with carefully selected design parameters, is an important first step. This program aids researchers in the design of primers for PCR by suggesting sequences for primers that are matched in their annealing temperatures (see Chapter 18) and will generate a PCR product of the desired size for the researcher. Further enhancements to this program have helped researchers to avoid some common primer-derived pitfalls in PCR that will be discussed in Chapter 18, such as primer dimers and hairpins that can result in there being no PCR products.

Primer-BLAST is a combination of Primer3 and BLAST to find primers for PCR and check that they are specific through BLAST. The BLAST part of the program uses the Primer3 output as the query sequence and finds any other hits in the targeted genome that mean the primers would generate a PCR product other than the one desired (**Figure 16.23**). Such non-specific PCR products can be a huge issue in PCR and waste time in the laboratory, therefore good primer design at the start can save aggravation later.

There are several programs that can help identify restriction enzyme digest sites within DNA from sequence data, which can aid in the design of experiments in the laboratory, such as the generation of recombinant DNA (see Chapter 18). Often the companies that supply restriction endonucleases have these tools on their web sites. For example, New England Biolabs has their NEBcutter to identify restriction sites in sequence data. These programs can often also suggest enzymes that have compatible sticky ends, facilitating the joining together of fragments of DNA (**Figure 16.24**).

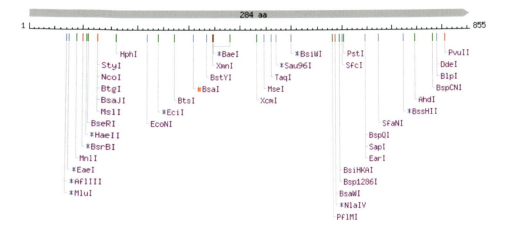

Figure 16.24 NEBcutter results. Using NEBcutter, the restriction enzyme recognition sequence sites within the *E. coli rpoH* gene sequence can be readily identified and displayed either graphically, as here, or in a list ordered by their position within the sequence.

Key points

- Base calling, assembly, and polishing of the genomic data from the sequencing machine all precedes identification of the genetic features in the DNA.

- Genome sequence annotation to find all of the CDSs, RNAs, and other genetic features often starts with an automated annotation pipeline.

- The NCBI Prokaryotic Genomes Automatic Annotation Pipeline, RAST, and Prokka are three popular automated bacterial genome annotation systems.

- All automated annotations should be manually curated, which can be facilitated through use of a visualization program, such as Artemis, GenDB, or Mauve.

- Comparisons between genome sequences can reveal important similarities and differences, which is why programs such as ACT and Mauve that can align multiple whole bacterial genomes are beneficial research tools.

- Galaxy provides a web-based interface for bioinformatics tools that are otherwise only available via command-line and tracks the workflow applied to data, making reporting of analysis steps in publications easier.

Terms, questions, and discussions

Key terms

Define the following terms introduced in this chapter. Check your answers using the definitions in the Glossary. These terms are also available as Flashcards on-line.

Base calling	Multi-locus sequence typing	Polishing
Clonal complexes	Orthologous regions	Reciprocal best hits

Self-study questions

Answer each question using 50–100 words or a table or labeled diagram. Advice on where to find answers to these questions is available on-line.

1. Outline the stages involved in assembly, which prepares genome sequence data for analysis.

2. What kinds of features can be identified in genome sequence data and annotated for inclusion with GenBank/ENA/DDBJ entries?

3. Download the complete genome sequence of any bacterial strain sequenced in the 1990s or early 2000s. Using the FASTA format of this data, generate a new automated annotation using PGAP, RAST, or Prokka. Note that this exercise will take some time to generate the annotations and if you choose to try Prokka, this requires local installations. Save this annotation.

4. Use the Artemis Genome Browser to view the annotation generated in question 3. Look through some of the genomic data and note at least three regions: one that looks like it could be an operon; one that includes rRNA or tRNA loci; and one where CDSs switch from one strand of the DNA to the other.

5. Use Artemis to generate a circular chromosome figure of the genome downloaded in question 3, displaying features from the annotation generated in question 4. Try to display the rRNA loci, %G+C, and GC skew. How does this compare to the original publication about this genome?

6. Use ACT to compare the genome sequence from question 3 to one or two other genome sequences from the same species. Are there any regions of rearrangement and/or inversion? Are there regions of difference, which are present in one genome sequence but not in the other?

7. Download and use Mauve to compare the original annotation of the bacterial genome sequence from question 3 to the automated annotation generated. Use progressiveMauve for the alignment. Scan through the annotations and note up to five differences between the annotations. Why do you think there are these differences?

8. Use progressiveMauve to compare the same genome sequences that were compared in question 6 with ACT. How are the differences between the genome sequences displayed in Mauve, including rearrangements, inversions, and regions of differences? Is anything notable in Mauve that was not in ACT?

9. Adjust the zoom and the settings in the Mauve alignment from question 8 to display a region of difference of your choice. Decide if you prefer the LCBs to be a solid color or to display the degrees of similarity by including white within the colored blocks. Export the image and save it.

10. Use the search function within Mauve to find *dnaA* in the genomes aligned in question 8. Zoom in to the DNA level and assess the sequence across all of the genome sequences to see if there are any differences in any of the aligned genome sequences.

11. Access Galaxy through the on-line web interface, register, and work through the Galaxy 101 tutorial. Note how the stages of analysis are recorded and saved, so they can be referred to later.

12. Using what you have learned from Galaxy 101, apply this knowledge to another sequence. This might be the sequence downloaded for use in question 3 or recently generated genome sequence data.

Discussion topics

These topics are presented for discussion in study groups, as part of class discussions, or on your own. These questions go beyond what is directly covered in this part of the book. Use the research literature and other reading to explore these topics in more depth. Tips to help prepare for topic discussions are available on-line.

13. Compare an early bacterial genome sequence publication, such as the first *Escherichia coli*, *Bacillus subtilis*, or *Haemophilus influenzae* genome sequence, to a recent genome sequencing publication. What are the striking differences about what is reported in these papers? What are the differences in the methods used in sequencing, assembly, annotation, and analysis?

14. Much of the bacterial genome sequence data that is generated is not assembled to a closed circular genome (for those species that have circular genomes). Discuss the reasons why this process is difficult and time consuming. What analyses can be done with an assembly that is still in several contigs (or scaffolds)? Which tools described in this chapter can be used on this kind of genomic data?

15. Discuss how MLST has contributed to understanding a specific bacterial population, paying particular attention to the methods used in the sequencing, sequence data assembly, annotation, and analysis. What insight was gained into the specific bacterial species, its geographic and/or chronological distribution, its antimicrobial resistance profile, and the ways it is/was transmitted?

Further reading

A few things happen to the genome sequencing data before the search for genes

Land M, Hauser L, Jun S-R, Nookaew I, Leuze MR, Ahn T-H *et al.* Insights from 20 years of bacterial genome sequencing. *Funct Integr Genomics*. 2015; 15(2): 141–61.

Automated annotation pipelines usefully combine feature identification tools

Brettin T, Davis JJ, Disz T, Edwards RA, Gerdes S, Olsen GJ *et al.* RASTtk: A modular and extensible implementation of the RAST algorithm for building custom annotation pipelines and annotating batches of genomes. *Sci Rep*. 2015; 5: 8365.

Delcher A, Bratke K, Powers E, Salzberg S. Identifying bacterial genes and endosymbiont DNA with Glimmer. *Bioinformatics* 2007; 23(6): 673–9.

Seemann T. Prokka: Rapid prokaryotic genome annotation. *Bioinformatics* 2014; 30(14): 2068–9.

Tatusova T, DiCuccio M, Badretdin A, Chetvernin V, Newrocki EP, Zaslavsky L, Lomsadze A, Pruitt KD, Borodovsky M, Ostell J. NCBI prokaryotic genome annotation pipeline. *Nucleic Acids Res*. 2016; 44(14): 6614–24.

Visualization of an automatically generated annotation can aid manual curation

Carver T, Harris SR, Berriman M, Parkhill J, McQuillan JA. Artemis: An integrated platform for visualization and analysis of high-throughput sequence-based experimental data. *Bioinformatics (Oxford, England)* 2012; 28(4): 464–9.

Meyer F, Goesmann A, McHardy AC, Bartels D, Bekel T, Clausen J *et al.* GenDB – an open source genome annotation system for prokaryotic genomes. *Nucleic Acids Res*. 2003; 31(8): 2187–95.

Spencer-Smith R, Roberts S, Gurung N, Snyder LAS. DNA uptake sequences in *Neisseria gonorrhoeae* as intrinsic transcriptional terminators and markers of horizontal gene transfer. *Microb Genomics*. 2016 Aug; 2(8): e000069.

Genomes can be aligned, just like genes can be aligned

Alikhan NF, Petty NK, Ben Zakour NL, Beatson SA. BLAST Ring Image Generator (BRIG): Simple prokaryote genome comparisons. *BMC Genomics* 2011; 12: 402.

Carver TJ, Rutherford KM, Berriman M, Rajandream MA, Barrell BG, Parkhill J. ACT: The Artemis comparison tool. *Bioinformatics* 2005 Aug 15; 21(16): 3422–3.

There is value in typing data, even in the genomics age

Falush D, Wirth T, Linz B, Pritchard JK, Stephens M, Kidd M *et al.* Traces of human migrations in *Helicobacter pylori* populations. *Science* 2003 Mar 7; 299(5612): 1582–5.

Matsunari O, Shiota S, Suzuki R, Watada M, Kinjo N, Murakami K, Fujioka T, Kinjo F, Yamaoka Y. Association between *Helicobacter pylori* virulence factors and gastroduodenal diseases in Okinawa, Japan. *J Clin Microbiol.* 2012 Mar; 50(3): 876–83.

Galaxy provides a full analysis suite for biological data

Afgan E, Baker D, Batut B, van den Beek M, Bouvier D, Čech M *et al.* The Galaxy platform for accessible, reproducible and collaborative biomedical analyses: 2018 update. *Nucleic Acids Res.* 2018 July 2; 46(W1): W537–44.

Laboratory Techniques

In previous chapters, the theory of laboratory techniques has been described. In this chapter, the application of those theories will be explored in detail, with discussion of the specifics of the techniques including tips and recommendations for protocols and experimental design, as well as troubleshooting and alternative strategies.[1,2]

The study of bacterial genetics and genomics fundamentally focuses on DNA, therefore starting with lysis of bacterial cells for DNA extraction

To extract DNA, several laboratory methods have been developed and improved upon. For most extraction methods, bacterial cells are suspended in lysis buffer. This is usually made up of TE, which is 10 mM Tris-HCl (tris hydrochloride) and 1 mM EDTA (ethylenediaminetetraacetic acid) to which is added a detergent, such as SDS (sodium dodecyl sulfate) at a concentration of 0.5%. Alternatively, 1% Triton X-100 can be used as the detergent. Detergent agents not only aid in disruption of the cell membranes, but they also remove membrane proteins and lipids from the DNA extraction.

For Gram-negative bacterial cells, add 200 µg/mL proteinase K to the lysis buffer and incubate at 56°C for 30 minutes to 1 hour. For Gram-positive bacterial cells, lysis is more difficult, largely due to the thickness of the peptidoglycan layer. The lysis of acid-fast bacteria can also be tricky. Addition of lysozyme can be beneficial in these cases, however lysozyme is less effective against staphylococci and many *Bacillus* strains. Lysozyme is especially effective when combined with EDTA, which is present already in TE (Tris-HCl and EDTA) and is effective at disrupting bacterial cell walls, provided the enzyme has access to the peptidoglycan layer, which it does not in Gram-negatives unless the outer membrane is otherwise compromised. Lysostaphin is effective at aiding in the lysis of staphylococci, but is only effective against staphylococci.

---

[1]This chapter is not a laboratory manual, however, several laboratory techniques are described in sufficient detail that they could be translated into laboratory use with minimal additional research. Such additional research must include completion of all locally required safety documentation, full awareness of the hazards involved in all experiments, and use of the appropriate personal protective equipment by all persons affected. Experiments should only be performed in an appropriate laboratory environment. The exception to this location restriction is the at-home activity described at the end of the chapter for extraction of DNA in the kitchen using common household items.

[2]Because next-generation sequencing protocols are specific to their sequencing machine platforms and are being continuously improved upon, they are not discussed in this chapter. Rather, there are overviews of these in Part IV of this book. Additionally, many manufacturers have specific kits for some of the techniques described here in this chapter that may have enhanced features or success rates. Familiarity with the techniques here will provide a basis for understanding the scientific basis of such kits and therefore aid in optimization or troubleshooting of such kits. Whether using a long-standing laboratory protocol or one from a purchased kit, always read all the way through a protocol before starting an experiment. Be sure you understand the experiment completely before beginning the experimental work and if not, that you research the steps that you do not understand. Knowing the purpose of all of the steps of a protocol will help with troubleshooting should you obtain results that are unexpected. Being familiar with a protocol before starting it will enable you to prepare all of the solutions, reagents, equipment, and materials that will be needed for the experiment, including setting temperatures for incubators or water baths or having bacterial cultures ready for the next experiment.

Often physical methods of breaking open Gram-positive cells work best, such as beating the cells in a tube with small sterile beads or **sonication**. Sonication applies sound to agitate the sample, thereby breaking open the bacterial cells and releasing the DNA. To be precise, this technique uses ultrasonic waves delivered via an ultrasonicating bath or ultrasonic probe. Spores can also be difficult to break open to be able to extract out the DNA. Like Gram-positive bacteria, physical methods are generally used to get the spores open.

DNA extraction using phenol produces very pure, large quantities of DNA

When a large quantity of very pure DNA is required, an old and established method of extraction using phenol is employed, although this carries with it hazards. Phenol is acutely toxic and causes chemical burns, while at the same time numbing the skin so that exposure is not immediately detected. It is imperative that material safety data sheets are read for this reagent before use, the risks are assessed, and gloves are changed frequently; phenol goes through nitrile gloves rapidly, therefore these should be avoided for phenol work.

There are phenol DNA extraction methods, and RNA extraction methods discussed later in this chapter, that have eliminated the lysis step by shaking the cells in phenol either on its own or with glass beads. Either way, the cells need to be broken open for the DNA to be released from the bacterial cells. The extraction method starts with the DNA in an aqueous solution.

Phenol/chloroform/isoamyl alcohol (25:24:1) pH 7.8–8.2 is used for DNA extraction. The pH is important. If acidic phenol is used, the DNA will not be recovered; the DNA will be in the phenolic layer and the RNA alone will be in the aqueous layer. Acidic phenol is used for RNA extraction, which will be covered in a later section of this chapter. Check if the supplied bottle of phenol/chloroform/isoamyl alcohol has a layer of water or buffer on the top. If so, do not remove this; pipette from below the water level to use the phenol/chloroform/isoamyl alcohol. Add an equal volume of phenol:chloroform:isoamyl alcohol to bacterial cells suspended in broth or PBS. Mix until a suspension is formed. Some protocols suggest vortexing this mixture for 1 minute, while others recommend gentle agitation for 10 minutes. The latter results in more intact DNA, whereas the former will lyse more cells. The goal of this step is to create an emulsion.

Phase separation and DNA precipitation in a phenol DNA extraction result in isolated DNA

Centrifuge the emulsion at the maximum speed allowable for the centrifuge and tubes for at least 1 minute to separate the aqueous and phenolic phases (**Figure 18.1**). This may require up to 10 minutes of centrifugation. Generally, the upper layer will be the aqueous phase and thus contain the DNA. However, if the sample is high in salt or otherwise alters the properties of the emulsion, the aqueous phase can be at the bottom. Using a colored phenol available from many suppliers can aid in the identification of the phenolic and aqueous phases. At this stage, the aqueous phase needs to be transferred to a new tube, however it is important to be careful not to disturb the interphase, the layer between the aqueous and phenolic phases (see Figure 18.1). It is also important not to disturb the phenol layer. The interphase contains proteins that will contaminate the DNA extraction, including potentially DNases that can degrade the DNA and other proteins that may inhibit downstream experiments. Phenol contamination of the DNA from the phenol phase at the bottom of the tube may also inhibit downstream experiments as well as being a hazard.

Transfer the aqueous phase to a new tube by pipetting with a 100–1,000 μL volume micropipette. Safely dispose of the phenolic phase, tubes, and tips in chemical waste according to hazardous waste protocols.

To precipitate the DNA out of solution, add 3 M NH_4OAc at one-tenth of the volume of recovered aqueous phase, pipetted under the liquid level of the aqueous

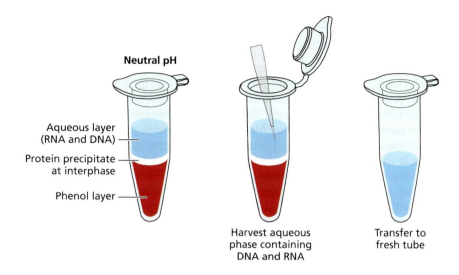

Neutral pH

Aqueous layer
(RNA and DNA)

Protein precipitate
at interphase

Phenol layer

Harvest aqueous
phase containing
DNA and RNA

Transfer to
fresh tube

Figure 18.1 Phenol extraction phases. Following centrifugation, the phases separate into a top aqueous phase containing DNA and RNA, a bottom phase that is the phenol, and a middle interphase layer of protein. The top layer is transferred via pipette to a fresh tube, taking care not to disturb the interphase or phenol layers.

phase in its new tube and mix (**Figure 18.2**). Next, add 2.5 volumes of 100% EtOH to the aqueous phase and mix (**Text Box 18.1**). At this stage, it should be possible to start to see the DNA floating in the mixture, appearing as a wisp that may have bubbles attached to it (see Figure 18.2). Extraction can continue from here or the sample can be chilled, either at −80°C for 1 hour or −20°C overnight, which is believed to increase yield. To pellet the DNA to the bottom of the tube, centrifuge at a safe maximum speed for 2–5 minutes, ideally at 4°C (see Figure 18.2). Carefully remove the ethanol supernatant from the DNA pellet. Wash the pellet with 70% EtOH by adding 200 μL to the pellet, repeating the centrifugation, and again carefully removing the ethanol supernatant from the DNA pellet. For increased purity, wash again with 70% EtOH.

The DNA pellet must be air dried, allowing the ethanol to evaporate. Leave the tube open on the bench in a clean location until all droplets are gone. To speed up this process, be sure all of the 70% EtOH liquid is removed carefully with a small pipette tip. DNA that is pure will be hard to see as a pellet. DNA that is contaminated with protein will appear as a white pellet, whereas pure DNA will be more transparent and can, at times, have a yellow tinge. Once the pellet is completely dry, suspend the DNA pellet in 10 mM Tris-EDTA pH 8.5 or ultra-pure water. Store the DNA at 4°C short term or at −20°C long term in aliquots to avoid repeated freeze–thaw. Freezing and thawing nucleic acids breaks the backbone, particularly of genomic DNA, resulting in fragmented DNA, therefore repeated freeze–thaws should be avoided.

Text Box 18.1

If you have 100 μL of aqueous phase, first add 10 μL of 3 M NH_4OAc and then 250 μL 100% EtOH.

To the aqueous phase, add sodium acetate under the surface of the liquid

Add ethanol to precipitate the DNA

After centrifugation, a pellet will be evident at the bottom of the tube

Figure 18.2 DNA precipitation. To the aqueous phase, add sodium acetate under the surface of the liquid. Then, add ethanol to precipitate the DNA. After centrifugation, a pellet will be evident at the bottom of the tube.

There are a few additional considerations to remember when doing restriction digestions

Add the restriction enzyme last. In fact, add all enzymes last to reactions. These are the active components, often stored at −20°C. Remove them just before use and put them right back at −20°C. There is no need to defrost enzymes. Enzymes are supplied in solutions containing glycerol, so they do not freeze. This also means that they are denser than water, therefore it is important to mix enzymes such as restriction endonucleases and polymerases added to PCR (later in this chapter) into reactions. Do so gently, stirring with the pipette tip, so as not to shear the DNA and damage it.

Incubate restriction digestions at the correct temperature, usually 37°C, for the correct time. The time for a digestion was usually 1 hour, however improvements in restriction digestion technology in 2006 reduced the time of the same reaction to 5 minutes for some restriction enzymes. Purchasing fast enzymes can be a time saving on protocols, enabling more research to be done in a day. Note that if the restriction enzyme being used has a warning about **star activity**, do not leave the reaction longer than needed to achieve digestion and do not use more enzyme than is needed to achieve digestion; leaving the reaction longer or using too much enzyme will make non-specific cuts.

Some enzymes can be **heat inactivated**, which stops the enzyme cutting DNA. This can be useful if, for example, the DNA needs to be digested with two different enzymes but there is no compatible buffer, so the two digestions must happen sequentially rather than together. Whether a specific enzyme can be heat inactivated and at what temperature is indicated by the supplier.

Restriction digestions are used to change DNA sequences and join sequences together

When joining together DNA sequences, such as when generating recombinant DNA, look for enzymes with **compatible cohesive ends**, if there isn't an enzyme that cuts both of the bits of DNA to be combined. For example, when wanting to join two pieces of DNA, it would be simple if both of them could be cut by the *Eco*RI restriction enzyme, since the sticky ends of the restriction fragments would hybridize together, facilitating ligation of the fragments. However, if there is no one enzyme that cuts both pieces of DNA, it may be possible to find two different enzymes that generate the same sticky ends. For example, *Eco*RI and *Mfe*I both generate TTAA 5′ overhangs (**Figure 18.6**). Once cut, the overhanging ends of the compatible cohesive ends will match (see Figure 18.6).

To prevent a plasmid from sticking back together spontaneously via hydrogen bonding between the bases after being cut with a single enzyme or enzymes with compatible ends, treat the plasmid with phosphatase. The phosphatase enzyme removes the terminal phosphate, preventing rejoining of the DNA by ligase enzymes (**Figure 18.7**). Alkaline phosphatase removes the 5′ phosphate groups from the plasmid vector so it will not ligate back together. In order for ligation to happen, the ligase enzyme needs to use the insert, the piece of DNA to be cloned into the plasmid, which has the 5′ phosphates. Alkaline phosphatase can be added to the restriction enzyme reaction when the plasmid vector is digested (see Figure 18.5); the water is reduced to allow for the added volume of the alkaline phosphatase

Figure 18.6 Compatible cohesive ends. The sequence shown on the top is cut by the restriction enzyme *Eco*RI, with the recognition sequence GAATTC. This cuts between the G and the A (staggered line), leaving a TTAA 5′ overhang, shown on the bottom strand. The sequence shown in the middle (in blue) is cut by the restriction enzyme *Mfe*I, with the recognition sequence CAATTG, which is slightly different. This cuts between the C and the A (staggered line), also leaving a TTAA 5′ overhang, shown on the bottom strand. Because both enzymes leave TTAA 5′ overhangs, they have compatible cohesive ends, which means that the fragments generated can be joined to one another, as shown in the bottom sequence. Here, the left part comes from the top sequence and the right (in blue) from the middle.

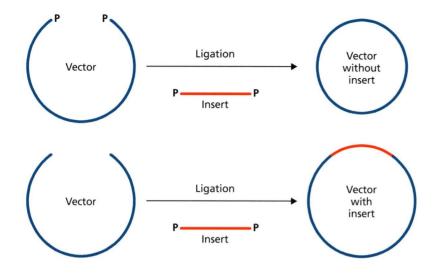

Figure 18.7 Alkaline phosphatase to prevent plasmid ligating back on itself. In a ligation of a vector and insert, the plasmid vector is most likely to ligate back together without the insert, if the ends of the vector are compatible (top panel). Alkaline phosphatase removes the 5′ phosphate from DNA (P). This phosphate is required for ligation. Using alkaline phosphatase on the plasmid vector prevents the plasmid from ligating back together. The plasmid will only circularize with the insert (red), which still has phosphate ends (bottom panel).

enzyme. Before proceeding to the next stage in cloning, the DNA fragments need to be purified away from the alkaline phosphatase, restriction enzyme(s), and buffers. This can be achieved using a variety of commercially available reaction clean-up kits.

If the cloning strategy means that there is a need to make blunt ends, but the available restriction digest recognition sites are sticky ends, it is possible to fill in the overhangs with the Klenow enzyme. Both 3′ overhang removal and fill in of 5′ overhangs are accomplished using the DNA polymerase I (Klenow) fragment (**Text Box 18.5**).

Text Box 18.5

To remove 3′ overhangs and fill in 5′ overhangs, add 1 unit Klenow per μg DNA with restriction enzyme buffer or ligation buffer and 33 mM of each dNTPs for 15 minutes at 25°C. To stop the reaction, add EDTA to a final 10 mM concentration and heat to 75°C for 20 minutes.

Cut ends of DNA need to be ligated together to complete cloning

To generate recombinant plasmids commonly used when cloning genes or sequences that will be manipulated or transformed into bacteria, it is necessary to ligate the insert into the plasmid. Ligation covalently connects the DNA backbone of the restriction enzyme digested DNA fragments. T4 DNA ligase joins the 5′-phosphate and the 3′-hydroxyl groups of adjacent nucleotides (**Figure 18.8**).

The ratio of insert to vector in a ligation reaction is important and an on-line calculator that takes account of the length of the fragments to be ligated is recommended. Otherwise, if the insert is smaller than the vector, a molar ratio of 3 insert to 1 vector is usually adequate. In total, approximately 100 ng of DNA should be in the ligation reaction (**Figure 18.9**). The ligation reaction should be set up on ice, adding the ligase enzyme last (**Text Box 18.6**). Chill the tube on ice and transform 1–5 μL of the mixture directly into 50 μL of competent cells.

Text Box 18.6

Generally, a 10 μL ligation reaction is set up with 1 μL of 10× ligase buffer, 25 ng vector DNA, 75 ng of insert DNA, enough dH$_2$O to take the final volume to 10 μL, and finally 1 μL DNA ligase (see Figure 18.9). Gently mix by stirring the reaction with the pipette tip. To ligate cohesive ends, incubate the ligation reaction overnight at 16°C or at room temperature for 10 minutes. To ligate blunt ends, or single-base overhangs, incubate at 16°C overnight or at room temperature for 2 hours. Alternatively, a thermocycler can be set to alternate between 30 seconds at 10°C and 30 seconds at 30°C. The ligase enzyme can be heat inactivated at 65°C for 10 minutes.

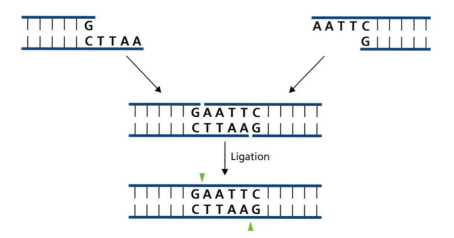

Figure 18.8 Ligation closes the backbone of DNA. Breaks in the backbone of DNA are joined by the ligase enzyme. For example, when two pieces of DNA cut by restriction enzymes come together with sticky ends, hydrogen bonding between the bases aligns the strands and keeps them together, but the backbone of the DNA is discontinuous. Ligase is able to join the 5′-phosphate and the 3′-hydroxyl groups of adjacent nucleotides, completing the helix.

Volume	Reagent
6 µL	dH₂O
1 µL	10x ligase buffer
1 µL	Vector DNA (25 ng/1 µL)
1 µL	Insert DNA (75 ng/1 µL)
1 µL	DNA ligase
Total 10.0 µL reaction	

Figure 18.9 Ligation reaction set-up.
To set up a typical ligation reaction, four things are needed: ligase; the buffer for the ligase; and the two pieces of DNA for the ligase to stick together. The reaction needs to be in the proper final volume, made up with distilled water (dH₂O). Start with the largest volume, pipette smaller volumes into it, and add the enzyme last, as shown here in this table.

The optimal temperature for T4 DNA ligase enzyme activity is 25°C, however, this is not necessarily the best temperature for annealing of the ends of the DNA fragments. Sticky end ligations have overhanging bases that can hydrogen bond and therefore efficiently ligate at room temperature, at about 22°C in 3 hours or 4–8°C overnight. Blunt end ligations are less efficient and generally are done at 15–20°C for 4–18 hours. If you are unsure how long to leave a ligation or at what temperature, it is possible to transform the ligation into competent bacterial cells after a few hours with 1 µL of the ligation reaction and then continue with the rest of the reaction overnight, if desired. There are suggestions that ligations done on melting ice can be very successful. On ice, the cold temperature permits the hybridization of the DNA fragments together. As the ice melts, the temperature rises to what is more optimal for the ligase, yet the temperature rises slowly so the hybridization is not lost.

Important considerations when performing ligations

Ligase buffer contains ATP and is stored frozen at −20°C. As with restriction endonuclease buffer, the ligase buffer must be thawed completely to preserve the correct ratios of the components of the buffer. It must not be used when partially thawed. The ATP in this buffer is an additional complication because it degrades with each freeze–thaw cycle. It is therefore recommended that the ligase buffer is completely defrosted when it arrives from the supplier and carefully aliquoted into smaller volumes of 5–10 µL. Since buffers are usually provided from suppliers in excess, a new aliquot tube can generally be used each time, preserving the activity of the ATP. The creation and use of aliquots also minimizes the degradation of DTT (dithiothreitol), which is present in the ligation buffer. DTT can precipitate upon freezing. Vortex the thawed ligation buffer for 1–2 minutes to resuspend the DTT prior to use.

It is very easy to neglect to do controls for things like ligations, but when the ligation does not work, a control can help you to work out why. In this case, it is important to include alongside the ligation reaction a second tube that has only vector, with no insert DNA. This will both confirm that the vector has completely digested and that the phosphatase treatment has worked (when needed). If this control generates ligated circular DNA and any colonies on transformation, it means the vector was either not fully cut or the phosphatase did not fully work.

PEG, polyethylene glycol, is sometimes used in ligation reactions as an agent that promotes "molecular crowding," which means it brings the fragments together and so is particularly good for blunt end ligations. It is best, if this is done, to use PEG that is pure, and to adjust the molar ratio and check the resulting ligations because **concatemerizations** can occur. This is when two or more inserts ligate together, generating an unexpected construct. Some protocols that recommend PEG will use PEG 8000 at 10% final reaction concentration in a ligation of 3–6 hours. Other protocols recommend using final concentrations of 5% PEG 4000 in blunt reactions. There are commercially available products that provide more rapid ligation than standard ligase and buffer. These tend to make use of optimized ligation buffers that include PEG. One manufacturer's quick ligation reaction buffer differs from their standard buffer through the addition of a final concentration of 6% PEG 6000, for example.

Cloning of sequences is often important in bacterial genetics and genomics research

To understand the functions of bacterial sequences, often these are cloned into plasmids or other vectors for manipulation. Plasmids engineered for molecular cloning have a **multiple cloning site (MCS)** with many restriction enzyme recognition sites. This is sometimes also called the **polylinker**. To make use of the plasmid for cloning, it is cut at the MCS using a restriction enzyme that either also cuts either side of the sequence to be inserted or that has compatible cohesive ends with the enzyme that will be used to cut the insert. As mentioned earlier, the addition of alkaline phosphatase to the restriction digestion of the plasmid will prevent the ligation reaction from just joining the plasmid back together. Instead, ligation to produce a closed circular piece of DNA will need to incorporate the desired insert. The sequence

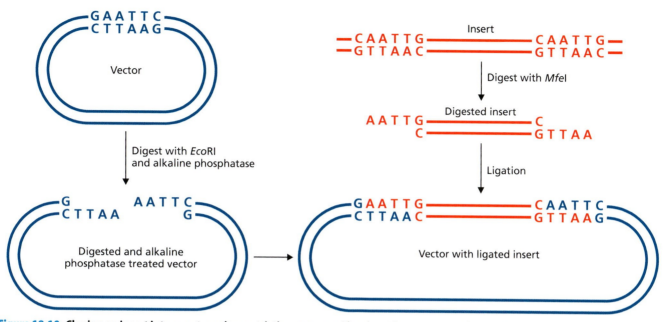

Figure 18.10 Cloning an insert into a vector using restriction enzymes. The plasmid vector is cut with a restriction enzyme, here *Eco*RI. Alkaline phosphatase is added to this reaction to prevent the plasmid ligating back upon itself. The insert is cut with *Mfe*I, which has compatible cohesive ends. The sticky ends of the insert and the vector join in the ligation reaction and the DNA backbone of the vector and insert are joined by ligase.

to be inserted into the plasmid is cut with either the same restriction enzyme as used to cut the plasmid or an enzyme with compatible cohesive ends (**Figure 18.10**).

With the preceding strategy, the insert can be cloned into the plasmid vector in either orientation, which may not matter depending on the experiment. However, sometimes **directional cloning** is desirable, where the insert needs to be in the plasmid in a specific orientation. In this case, the cloning is designed so that the plasmid is cut at the MCS with two different restriction enzymes that have incompatible cohesive ends (**Figure 18.11**). This means that the insert will only go into the plasmid in one orientation and it prevents the plasmid from being able to

Figure 18.11 Directional cloning of an insert into a vector using restriction enzymes. The plasmid vector is cut with two restriction enzymes, here *Eco*RI and *Bam*HI. Alkaline phosphatase is added to this reaction to prevent the plasmid ligating back upon itself. The insert is also cut with *Eco*RI and *Bam*HI, which means that the sticky ends of the insert and the vector join in only one orientation. During the ligation reaction, the DNA backbone of the vector and insert are joined by ligase.

be ligated back together because the two ends of the plasmid will not be compatible with one another. It is, however, still a good idea to include alkaline phosphatase here, because any incomplete digestions that only cut once will be prevented from upsetting the cloning by re-ligating. The tiny piece of DNA that is removed by the cutting with the two enzymes also needs to be purified away so it does not interfere in the ligation process. This can be done using a clean-up column or by resolving the restriction digestion band on an agarose gel and isolating the DNA band from the gel. Either would generally be done with a column-based kit. The sequence to be inserted is cut either with the same two restriction enzymes or with enzymes that have compatible cohesive ends to those used for the plasmid.

Selection of the restriction enzymes to be used for the cloning can be aided by on-line tools. It may be necessary to use blunt cutting enzymes, but these are less efficient in cloning because the ends of the DNA will not make use of the hydrogen bonding between nucleotide bases to aid the ligation process. It is possible to engineer restriction digest sites onto the ends of the sequence to be inserted, if necessary. This can be helpful if none exist, to place restriction recognition sites where they are needed, or to make use of restriction endonucleases that are the best choice for the cloning. As in the PCR description that follows, primers can be designed with additional sequence information at the 5′ end, which then becomes part of the amplified DNA.

TA cloning exploits a feature of PCR to rapidly clone sequences

One of the most popular methods of cloning, TA cloning, uses a PCR-generated product and inserts it into a plasmid vector by virtue of T and A overhangs, thus its name. TA cloning is simpler and quicker than traditional cloning with restriction enzymes because it is not reliant on there being convenient restriction enzyme recognition sites in the sequence or engineering them by PCR. However, TA cloning is not directional, as can be achieved with careful design of a traditional cloning experiment.

Taq and other PCR enzymes lack 5′ to 3′ proofreading activity and as such add an additional adenosine triphosphate to the 3′ end of the product (**Figure 18.12**). Because of the 3′ A overhangs on PCR products, these pieces of DNA can be cloned into plasmids with complementary 3′ T overhangs (see Figure 18.12). The process can be optimized by designing the PCR product so that primers have G's at their 5′ ends; this increases the probability of addition of the A. Choice of polymerase for the PCR is also important; not all polymerases will add the A's, including most with proofreading activity, which tend to be preferred for making accurate PCR products with no introduced mistakes in the sequence. To ensure accuracy, it is possible to generate the amplicon with a proofreading polymerase and then to add some *Taq* polymerase to the end of a proofreading PCR just to have it add on the A's or to purchase a special proofreader that adds on the final A to the product.

The vector can be purchased in a TA cloning kit or created in the laboratory. To generate a TA cloning vector, the plasmid is linearized with a blunt cutting restriction enzyme and then 3′ T's are added by terminal transferase from ddTTP (dideoxythymidine triphosphate). This is the same kind of T used in Sanger sequencing, which when incorporated will not allow another base to be added to it. In this way, only one T is added to each 3′ end of the cut plasmid (see Figure 18.12).

Some commercially available kits augment ligation and cloning with accessory proteins and exploitation of other systems

Fragments can be cloned into vectors without DNA ligase using TOPO TA cloning, where the process of TA cloning is enhanced by using a TA cloning vector with DNA topoisomerase I attached (**Figure 18.13**). Topoisomerase acts in replication to cleave

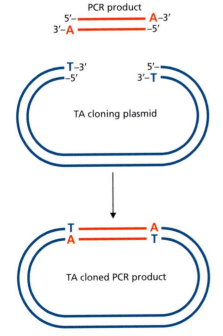

Figure 18.12 TA cloning. PCR products have 3′ A overhangs (top), therefore it is possible to clone them into plasmids that have 3′ T overhangs (middle). The A ends on the PCR product hydrogen bond to the T ends on the plasmid, circularizing the plasmid with the PCR product as an insert (bottom), resulting in a TA cloned PCR product in the plasmid vector.

Figure 18.13 TOPO TA Cloning. TA cloning (see Figure 18.12) through the use of a TOPO TA cloning vector, a plasmid that has 3′ T overhangs like a TA cloning vector and also has DNA topoisomerase I attached to these ends. When a PCR product with 3′ A ends hybridizes with the plasmid, the topoisomerase joins the DNA backbones, making TOPO TA cloning more efficient than TA cloning.

and rejoin supercoiled DNA. TOPO vectors have topoisomerase covalently bound to the cut ends of the plasmid. During cloning, the PCR product end associates with the vector end and topoisomerase covalently links their backbones together. Some TOPO vectors also clone in PCR products made with proofreading enzymes, which lack the additional A required for TA cloning. These blunt TOPO cloning reactions are made more efficient by the topoisomerase bound to the vector than would be a traditional blunt ligation.

There are other cloning methods that rely on DNA pairing and homologous recombination, needing compatible regions between vectors and DNA inserts. For example, ligation independent cloning (LIC) uses long overlaps between insert and vector to stabilize the hybridization of base pairs compared to traditional restriction enzyme cloning where this would be generally no more than 4 bp. Overhangs of 10–12 bp are made by T4 DNA polymerase and specifically designed primer sequences. The insert and vectors annealed together contain single-stranded nicks that are repaired once the plasmids are transformed into the *Escherichia coli* where they are replicated.

Gateway cloning is a proprietary system of recombination-based cloning owned by Invitrogen and designed to enable researchers to easily move a cloned piece of DNA between different plasmids. Each move retains the reading frame of the insert sequence by use of a set of recombination sequences, the Gateway *att* sites and system-specific cloning enzymes. To use the system, the insert first goes into the Gateway entry clone by adding Gateway *attB1* 5′ and Gateway *attB2* 3′ via engineered PCR primers. The amplicon with *attB1* and *attB2* ends is recombined into the Gateway donor vector by a proprietary enzyme mix, creating the Gateway entry clone. The cloned insert can then go into Gateway destination vectors for a variety of experiments. Gateway is based on site-specific recombination of λ bacteriophage, which enables the prophage genome to integrate into the *E. coli* chromosome. The λ bacteriophage genome has a recombination site *attP* and *E. coli* has a corresponding recombination site *attB*. Two enzymes are involved in the prophage integration, integrase from λ bacteriophage and IHF from *E. coli*. Integration creates *attL* and *attR* sites flanking the prophage that enable reversible excision of the prophage from the *E. coli* chromosome by the Xis phage protein. Gateway and other similar recombination-based cloning techniques are based on bacteriophage integration and excision mechanisms, which are covered in more detail in Chapter 21.

CPEC, circular polymerase extension cloning, uses polymerase to extend overlapping insert and vector sequences to make complete circular plasmids. The overlaps hybridize and act as primers for polymerase extension around the circular plasmid (**Figure 18.14**). As with LIC, the plasmids are transformed into the bacterial cells where the nicks are resolved during replication. It is possible with this process to include more than one overlapping insert in the design for CPEC in a design process similar to overlapping PCR, discussed later in this chapter.

Figure 18.14 Circular polymerase extension cloning, CPEC. In this cloning process, polymerase is used to complete the synthesis of the second strand of DNA around a single strand of plasmid and a single strand of insert, thus joining them together. A denatured single-stranded insert with overlapping similarity to a denatured single-stranded plasmid vector hybridize (left). The overlapping regions are used by polymerase to prime extension and generate the complementary strand (right).

Biotechnology

The use of biological processes or systems, such as the use of biological organisms, for the benefit of improving human life is **biotechnology**. This benefit can be in the broadest of senses. Biotechnology can clearly be the use of organisms to create or produce medicines, but it can also be to clean up our environment so that we have a healthier planet on which to live. Biotechnology can also be applied to the creation of novel cosmetic products that enrich our lives and make us happier, thereby bringing us joy, or which improve efficiency in industry that enables processes to be safer for human workers or for workers to be more productive, perhaps improving work–life balance. The use of microorganisms, especially bacteria, is a major part of the field of biotechnology. While many biological systems are used in biotechnology, this chapter, in keeping with the theme of this book as a whole, will focus only on those that use bacteria in some way to achieve their end result.

Biotechnology is far older than genetic engineering

When we think of biotechnology, we tend to think of modern scientific technologies. However, the use of biologicals for human benefit is ancient, if we consider processes such as beer brewing, mead and wine making, and cheese making. While the fermentation processes require yeasts, which are microbes outside the scope of this book, cheese making uses enzymes that originally came from sources such as calves' stomachs. These enzymes often now come from bioengineered bacteria, being recombinantly generated. Although cheese making itself predates recorded history, today we are able to make it using modern scientific techniques, including recombinant enzymes expressed in bacteria.

Rennet used in making cheese is extracted from the fourth stomach of calves that have been slaughtered for veal. After processing, rennet contains chymosin, a protease, as well as pepsin and lipase. The role of the chymosin enzyme is to coagulate milk, which is beneficial to calf digestion and enables humans to manufacture cheese. In older cows, the levels of chymosin drops, making the extraction from young calves' stomachs important for cheese making. Due to the expense and limited availability of rennet, cloning of the chymosin gene into bacteria and expression of the protein was an attractive solution. The fermentation-produced chymosin (FPC) from the bacteria is the same as the chymosin produced in the calf stomach. Today, most FPC is produced in either the fungus *Aspergillus niger* or the yeast *Kluyveromyces lactis*, but some is also made in *Escherichia coli*.

With the rise in awareness of lactose intolerance in humans, particularly adult humans, there has been an increasing market demand for lactose-free products. In infants, the lactase enzyme is made in the small intestine to digest human breast milk from the mother, but with age the expression of this enzyme can decrease or stop all together. Consumption of lactose-containing foods by lactose-intolerant individuals results in abdominal cramps, gas, diarrhea, and discomfort. In addition to non-dairy alternatives to milk and cheese products, biotechnology has come up with some solutions. The lactase enzyme has been produced using recombinant technologies similarly to chymosin and added in the production of milk and cheese products to reduce lactose, often to levels where companies are labeling products as lactose-free.

Biotechnology impacts many aspects of our lives and of research

Biotechnology is a broad discipline that touches many aspects of science, technology, economics, and society. In medicine, biotechnology is used to develop new pharmaceuticals such as monoclonal antibody-based therapies and recombinant protein-based drugs. In agriculture and environmental sciences, biotechnology is involved in genetic modification of plants, often involving bacteria as intermediates within which to manipulate the genetic material, as well as aiding in the development of efficient biofuels and use of bacterial species for bioremediation. In industrial biotechnology, bacteria aid in production of enzymes that enhance products as diverse as food and laundry soap.

Biotechnology permeates our lives, being present in our homes and market shelves. Biological laundry soaps contain digestive enzymes, generally lipases, proteases, or both. Some products, including dishwasher detergents, also contain amylases. Lipase is able to digest fats and oils while protease is able to digest proteins and starch, thereby removing the two most common sources of stains on clothing. Cloning and expression of lipase, protease, and amylase enzymes in bacteria using biotechnology means that large quantities of the enzymes can be made inexpensively. Indeed, around 25%–30% of enzymes are used by the detergent industry.

There is a fascinating ecosystem at the floor of the ocean that occurs when a whale dies and sinks to the deep ocean, below 1,000 m, called whale fall (**Figure 19.1**). This environment has created an ideal opportunity to explore microorganisms that are adapted for growth in cold water, less than 20°C, and that possess cold water adapted enzymes that function at these temperatures. For manufacturers of laundry soap, this is an ideal environment, especially with the move to set washing machines to cold water washes to save on energy and preserve clothing. From studies of whale fall, some of the lipase fat-dissolving enzymes in laundry soap now come from organisms that have evolved to live in cold water, making them ideal for cold water wash conditions (see Figure 19.1).

Large quantities of bacteria are grown in bioreactors to yield large quantities of recombinant proteins

To harvest large quantities of recombinant proteins for biotechnology applications, **bioreactor** systems are frequently used. Bioreactors are culture vessels intended for industrial use or for highly controlled long-term laboratory experiments. These closed bacterial growth systems enable large quantities of bacteria to be grown, enabling recombinant protein harvesting on a much larger scale than would be achieved in a typical laboratory bacterial culture (**Figure 19.2**).

At its most basic, a bioreactor is any engineered system that is able to support growth of microbes or biochemical reactions. Bioreactors can be vessels of less than a Liter or of 50,000 Liters, potentially more. Growth conditions within the bioreactor are carefully controlled and maintained, with new media added, temperature and pH controlled, oxygen levels adjustable, and the system, including stirring rate of the culture, fully monitored (see Figure 19.2). When growing large volumes of bacteria,

Figure 19.1 Whale fall ecosystems are sources of cold adapted enzymes. When a whale dies and its carcass falls to the deep ocean floor, it creates an ecosystem. Other organisms thrive on breaking down the nutrients of the whale body. These include organisms producing lipases adapted to function well at cold water temperatures. These have been collected, grown in laboratories, and used as enzymes in laundry soaps to break down fats in cold water laundry washes.

Whale fall on ocean floor

Organisms producing lipases

Enzymes in laundry soap

Figure 19.2 An example of a bioreactor. Bioreactors are contained systems for growth of bacteria under highly controlled conditions that can be carefully monitored. The culture is often stirred and aerated, as needed for the species, and temperature controlled, with the aid of the thermal jacket on the bioreactor vessel. As nutrients are depleted, more media can be added and waste removed.

such as *E. coli* and *Bacillus* spp. expressing recombinant proteins, the bioreactor maintains the optimal growth conditions to produce the best yield of protein. The bioreactor simulates the ideal conditions for the bacteria to grow and express the recombinant protein.

A typical liquid culture in a microbiology laboratory uses a broth medium in a shaking flask at a set temperature in an incubator. Under these conditions the temperature within the culture is not uniform, with the temperature within the flask differing from that toward the surface of the flask. The culture is shaken in an attempt to aerate the broth, however this also is not uniform throughout. A bioreactor attempts to address this and other variations in culture conditions to produce more reproducible cultures each time, as well as optimal conditions for each bacterial cell within the culture. However, setting up bioreactor cultures is more complex than a simple liquid culture, which is why bioreactors are not used for routine bacterial growth.

Human insulin expressed in *Escherichia coli* is a classic example of biotechnology

Insulin is needed for patients with diabetes. Insulin was once isolated from the pancreas of pigs, cows, and sometimes salmon. This was a slow and expensive process that was unable to keep up with the demands of clinical need in a growing population. In addition, some patients were allergic to insulin harvested from animals and others were not comfortable with use of medicines derived from animals. Overcoming these issues, the gene for human insulin was cloned into bacteria and expressed as a recombinant protein.

Humulin was licensed for human use in 1982. Two DNA sequences were introduced into *E. coli*: one for the insulin A chain of 21 amino acids and one for the B chain of 30 amino acids. In humans, insulin is made as one polypeptide chain, proinsulin, which has the A chain and B chain separated by the 33 amino acid C peptide (**Figure 19.3**). Proteolytic cleavage removes the C peptide. The A chain and B chain are then connected by two disulfide bonds; a third disulfide bond occurs within the A chain. Production of the A and B chain peptides in *E. coli* results in recombinant human insulin with the same structure (see Figure 19.3).

Today, millions of diabetics use synthetic insulin that is made mostly by bacteria or yeast in some cases. Recombinant insulin can be a perfect match for human insulin or it can be changed slightly using genetic engineering to produce insulin analogs, including insulin lispro, known as Humalog. Humalog was the first insulin analog, entering the market in 1996, and is a fast-acting insulin. The amino acid sequence of the B chain of Humalog differs from normal human insulin at the C-terminus, where the lysine and proline residues are reversed, which blocks the formation of insulin dimers and hexamers. Because these multimeric forms of insulin cannot form, there is more insulin available for fast absorption by the patient. This minor change gave patients more flexibility in managing their lives because they did not have to wait as

Shigella spp.

Clostridium spp.

Mycoplasma pneumoniae
Mycoplasma genitalium
Mycoplasma hominis
Mycoplasma fermentans

Ureaplasma

Chlamydia trachomatis
Chlamydia pneumoniae

Neisseria gonorrhoeae

Streptococcus pneumoniae
Streptococcus pyogenes

Haemophilus influenzae
Haemophilus parainfluenzae

Listeria monocytogenes

Campylobacter jejuni

Helicobacter pylori

Staphylococcus aureus
Staphylococcus epidermidis

Legionella pneumophila

Enterococcus faecalis

Figure 20.2 Bacterial species susceptible to Solithromycin (CEM-101). Solithromycin is a macrolide of the subclass fluoroketolide, which inhibits translation through its action on 23S rRNA.

this has meant that antimicrobials that are deemed too specific in their activity have, unfortunately, not been supported for development. This restriction could perhaps be overcome in the future by advances in rapid diagnostics, enabling personalized infection identification and treatment. Second, when key genes have been identified as targets for drug development, small-molecule drugs proved ineffective, even in the short term. Small-molecule interference with enzymes resulted in rapid selection for resistance, which does not make them viable for clinical use either. This limitation may be able to be overcome in the future through design of molecular inhibitors that contact the target enzyme at more than one site, more like current antimicrobials.

Extensive understanding of bacterial pathogen genes and the antibiotics that have been used against them for decades has led to the development of some novel antibiotics. Unfortunately, three that were introduced in recent years were discontinued due to serious adverse events: Telithromycin; Temafloxacin; and Trovofloxacin. Also introduced in recent years was Fidaxomicin, available specifically for *Clostridioides difficile* colitis, despite issues with antibiotics for single species. Three antibiotics are available for community-acquired pneumonia and acute skin infections: Ceftaroline; Oritavancin; and Telavancin. A further two, Dalbavancin and Tedizolid, are also available for acute skin infections. Each of these faces market pressures from existing antibiotics, which are less expensive, however they provide alternatives when there is resistance to the existing repertoire of antibiotics. There are also several antibiotics in clinical trials. Solithromycin (CEM-101) is a fluoroketolide, a subclass of macrolides acting on the 23S rRNA subunit to inhibit translation. This fluoroketolide inhibits 50S ribosomal subunit formation and causes errors in translation. The structural modifications of Solithromycin increase its ribosomal binding and reduce its likelihood of development of resistance compared to macrolides. It has been demonstrated to act against many bacterial species (**Figure 20.2**). Zoliflodacin (AZD0914) is a spiropyrimidinetrione topoisomerase inhibitor that inhibits DNA biosynthesis and causes double-strand DNA breaks to accumulate in the bacterial cell. Although the target is similar to fluoroquinolone antibiotics, the mechanism of action of Zoliflodacin is different. It has been shown to act against several bacterial species (**Figure 20.3**). Gepotidacin (GSK2140944) also inhibits DNA replication like fluoroquinolones, but also via a different mechanism. It has demonstrated antimicrobial activity against some species as well (**Figure 20.4**).

Staphylococcus aureus
Staphylococcus epidermidis

Streptococcus pneumoniae
Streptococcus pyogenes
Streptococcus agalactiae

Haemophilus influenzae

Neisseria gonorrhoeae

Legionella pneumophila

Clostridium difficile

Chlamydia pneumoniae
Chlamydia trachomatis

Mycoplasma genitalium

Figure 20.3 Bacterial species susceptible to Zoliflodacin (AZD0914). Zoliflodacin inhibits topoisomerase and thus inhibits the biosynthesis of DNA. It is a spiropyrimidinetrione antibiotic.

Staphylococcus aureus

Streptococcus pneumoniae
Streptococcus pyogenes

Escherichia coli

Moraxella catarrhalis

Haemophilus influenzae

Clostridium perfringens

Shigella spp.

Figure 20.4 Bacterial species susceptible to Gepotidacin (GSK2140944). Gepotidacin inhibits DNA replication.

Genomics can aid in the search for new antibiotics

The majority of antibiotics originally came from or were derived from antimicrobial agents produced by bacteria and fungi (**Figure 20.5**). These agents originally evolved to rid the organisms of competitors in their niche and there are many more undiscovered antimicrobial compounds made by the microbial world. However, most of the bacterial species cannot yet be cultivated in the laboratory because the correct media and growth conditions are not known. Fortunately, metagenomics sequences can reveal the genes within unculturable bacterial and fungal species, including potential new antibiotics.

Several undertakings have been initiated to isolate bacteria from soil in an attempt to identify bacteria producing novel antimicrobials. These include major research projects as well as school projects. Some focus on simple microbiology techniques, while others also use genome sequencing methods. For example, whole genome sequencing of *Pseudomonas* species from around plant roots has been used to identify novel antimicrobials made by these bacteria, which evolved against species in this niche but may be of use in combating human infections.

Researchers are also exploring unique environments for antibiotic producing bacteria, including those in the Antarctic and Atacama Desert in Chile. These sorts of investigations are also revealing that antibiotic resistance genes have been around far longer than our use of antibiotics. A β-lactamase gene within a metagenomics sequence was isolated from deep-sea sediment dated 10,000 years ago, reinforcing that these genes were present naturally as defenses against the antibiotics produced by other microbes. In addition to studies looking at the age of β-lactamase genes, others have used metagenomics to look at their diversity, discovering that it is much larger than previously believed, which may have implications for clinical control of infectious diseases with β-lactam-based antimicrobials.

Some old drugs are getting a new lease of life due to greater depth of understanding

β-**Lactam** antibiotics, such as penicillin, have been compromised in their usefulness by the acquisition of many bacterial species of β-**lactamases** conferring resistance. The β-lactam antibiotics are characterized by their ring structure, which is cleaved by the action of the β-lactamase enzyme (**Figure 20.6**). Understanding the horizontal transfer of the β-lactamase genes and analysis of the types of β-lactamases has led to developments of β-lactamase inhibitors that enable β-lactams to be used. The first of

(a)   **(b)**

β-lactam ring

β-lactam cleavage site

Penicillins

Figure 20.6 The β-lactam ring and its cleavage by β-lactamase enzymes. Characteristic of β-lactam antibiotics such as penicillins is the presence of the β-lactam ring (red). This is the target of the β-lactamase enzyme for the inactivation of this class of antibiotics. β-Lactamase enzymes cleave the bond in the ring between the nitrogen and the carbon that is bound to the oxygen.

Figure 20.5 Sources of antibiotics: bacteria and fungi. Some bacteria, such as *Streptomyces coelicolor* (a), and fungi, such as *Penicillium chrysogenum* (b), produce antibiotics as defense mechanisms against other organisms in their environment. In panel (a), the antibiotic undecylprodigiosin is seen as the red areas diffusing into the medium. In panel (b), the spherical droplets of yellow on the surface are penicillin G. In some cases, antibiotics produced by organisms such as these have been exploited by humans, either in their natural form or as derivatives of the natural product. (Photographs from Dr. Jeremy Burgess and Daniela Beckmann, respectively, Science Photo Library.)

Putting discoveries into practice

This chapter has presented and explored some novel ideas for alternatives to antibiotics and new ideas for controlling infectious diseases. Many of these are being put into practice and are in trials to be used in the near future. Included among these are bacteriophage therapies, which will be explored in the final chapter of this book. Research in the field of infectious disease control is advancing rapidly, motivated by the urgent need to address the growing antimicrobial resistance crisis, which not only impacts the healthcare of humans, but also the agricultural industry and veterinary medicine as well.

Key points

- New antimicrobials can be revealed through the study of bacterial genomes and understanding the role and function of essential genes and virulence genes.

- Metagenomics can aid in the search for new antibiotics through the identification of genes in bacterial species that cannot yet be cultured in laboratories.

- Pairing of β-lactam antibiotics and cephalosporin antibiotics with inhibitors of β-lactamases and extended spectrum β-lactamases can overcome some types of antimicrobial resistance.

- Reverse vaccinology has used genomic sequence data as the basis for the development of new vaccines, where antigens are predicted from the genomic information.

- Novel approaches to control infectious diseases include development of anti-virulence agents that prevent bacteria from causing disease.

- The resolution of genome sequence data provides details for epidemiology studies that can discriminate between lineages causing different diseases, identify transmission events, and trace the sources of infections so they can be contained.

- Surveillance of infections within hospitals is important for infection control; application of genome sequencing is improving rapid identification of outbreaks.

- Genome sequencing is changing our understanding of the importance of certain genes in disease, including revealing that the bacterial capsule is not necessary for species like *S. pyogenes* and *N. meningitidis* to cause infections.

Terms, questions, and discussions

Key terms

Define the following terms introduced in this chapter. Check your answers using the definitions in the Glossary. These terms are also available as Flashcards on-line.

β-Lactam	Broad-spectrum antibiotic	Extended spectrum β-lactamases
β-Lactamases	Cephalosporins	Nosocomial infections
Antivirulence agent	Epidemiology	Narrow-spectrum antibiotic
Bactericidal	ESKAPE pathogens	Surveillance
Bacteriostatic	Etiological agent	Zoonotic disease

Self-study questions

Answer each question using 50–100 words or a table or labeled diagram. Advice on where to find answers to these questions is available on-line.

1. When were most antibiotics discovered? What are some of the reasons for antibiotic resistance to these drugs?

2. What is the difference between a narrow-spectrum antibiotic and a broad-spectrum antibiotic? Will there be a different outcome for the patient if both are equally able to kill the bacteria causing disease?

3. What are the sources of most antibiotics? Give some examples of natural antibiotic producers. Why do they make antibiotics?

4. Have antibiotic resistance genes been around longer than we have been using antibiotics? Explain your answer.

5. Give examples of how combining two drugs, an antibiotic and an inhibitor, is being used to tackle antibiotic resistance. How do these drugs work together?

6. Briefly explain the processes and principles involved in reverse vaccinology.

7. What is the difference between a drug that is an antivirulence agent and one that is an antibiotic? Could they work together? What is the advantage of targeting virulence?

8. Name two targets of antivirulence agents that have been under investigation for development. How does blocking these targets aid in controlling infectious diseases? Are these antivirulence agents broad spectrum?

9. Give some examples of how our understanding of bacterial pathogens has been changed through analysis of bacterial genome sequence data.

10. Describe how genome sequencing can be used in epidemiology.

11. What does infection surveillance involve in a hospital and how can sequencing be used to help?

12. What features of the genus *Neisseria* facilitate exchange of genes between species? How can genome sequencing clarify the identity of what may appear to be *N. gonorrhoeae* based on disease symptoms and isolation location, but is actually *N. meningitidis*?

Discussion topics

These topics are presented for discussion in study groups, as part of class discussions, or on your own. These questions go beyond what is directly covered in this part of the book. Use the research literature and other reading to explore these topics in more depth. Tips to help prepare for topic discussions are available on-line.

13. Explore how monoclonal antibodies are being used to tackle issues in human health, specifically those related to a specific infectious disease. Is the infection itself being targeted or is it a toxin or other virulence factor? What other treatments do patients undergoing the monoclonal antibody therapy require? How is their outcome improved through the additional use of the monoclonal antibodies?

14. As mentioned in this chapter, azithromycin does not kill *P. aeruginosa*, but it does interfere with its quorum sensing. How can azithromycin be exploited as an avenue to combat *P. aeruginosa* infections through quorum sensing interference? Investigate studies in the research literature that have looked at this specifically or at biofilm and quorum sensing disruption in general. Discuss the advantages of repurposing existing antibiotics, rather than trying to find new antibiotics.

15. Explore developments in antivirulence drugs or monoclonal antibody therapies. How effective have these been at demonstrating their potential to control infectious diseases? Have there been issues with resistance as there have been with antibiotics?

Further reading

Some old drugs are getting a new lease of life due to greater depth of understanding

van Duin D, Bonomo RA. Ceftazidime/Avibactam and Ceftolozane/Tazobactam: Second-generation β-lactam/β-lactamase inhibitor combinations. *Clin Infect Dis*. 15 July 2016; 63(2): 234–41.

Bacterial genomics has led to the development of new vaccines

Serruto D, Bottomley MJ, Ram S, Giuliani MM, Rappuoli R. The new multicomponent vaccine against meningococcal serogroup B, 4CMenB: Immunological, functional and structural characterization of the antigens. *Vaccine* 2012; 30: B87–97.

Reverse vaccinology is providing leads for several bacterial diseases

Chiang MH, Sung WC, Lien SP, Chen YZ, Lo AFY, Huang JH *et al.* Identification of novel vaccine candidates against *Acinetobacter baumannii* using reverse vaccinology. *Hum Vaccines Immunother.* 2015; 11: 1065–73.

Maione D, Margarit I, Rinaudo CD, Masignani V, Mora M, Scarselli M *et al.* Identification of a universal Group B *Streptococcus* vaccine by multiple genome screen. *Science* 2005; 309: 148–50.

Meunier M, Guyard-Nicodème M, Hirchaud E, Parra A, Chemaly M, Dory D. Identification of novel vaccine candidates against *Campylobacter* through reverse vaccinology. *J Immunol Res.* 2016; 2016: 5715790.

New drugs are being developed that will contain the virulence of bacteria

Parrino B, Schillaci D, Carnevale I, Giovannetti E, Diana P, Cirrincione G, Cascioferro S. Synthetic small molecules as anti-biofilm agents in the struggle against antibiotic resistance. *Eur J Med Chem.* 2019 January 1; 161: 154–78.

Gene sequencing and genome sequencing improves the resolution of epidemiology of bacterial infectious diseases

Knight DR, Elliott B, Chang BJ, Perkins TT, Riley TV. Diversity and evolution in the genome of *Clostridium difficile*. *Clin Microbiol Rev.* 2015 Jul; 28(3): 721–41.

Genome sequencing can improve infection control for surgical site infections

Harris SR, Cartwright EJ, Török ME, Holden MT, Brown NM, Ogilvy-Stuart AL *et al.* Whole-genome sequencing for analysis of an outbreak of methicillin-resistant *Staphylococcus aureus*: A descriptive study. *Lancet Infect Dis.* 2013 Feb; 13(2): 130–6.

Köser CU, Holden MT, Ellington MJ, Cartwright EJ, Brown NM, Ogilvy-Stuart AL *et al.* Rapid whole-genome sequencing for investigation of a neonatal MRSA outbreak. *N Engl J Med.* 2012 Jun 14; 366(24): 2267–75.

Peacock SJ, Parkhill J, Brown NM. Changing the paradigm for hospital outbreak detection by leading with genomic surveillance of nosocomial pathogens. *Microbiology.* 2018 Oct; 164(10): 1213–9.

Bacteriophages

Bacteriophages are abundant and ubiquitous, their nucleic acids are the most numerous on Earth, and they are prevalent across the globe. They are diverse in their size, shape, and structure, as well as in the organization of the genes in their genome. Bacteriophages are non-motile, relying on sheer numbers and chance to reach their targets on bacterial cells, which is necessary for their replication.

Bacteriophages have been studied for just over 100 years

Bacteriophages were first discovered in 1915, by Frederick Twort, who recognized that bacteriophages inhibit bacterial growth. It was two years later, in 1917, that Félix D'Hérelle isolated bacteriophages and used them to prevent *Salmonella* infection of chickens. There are believed to be 10^{31} bacteriophages on the planet; that is 10 nonillion, enough for 10 bacteriophages for each bacterial or archaeal cell on Earth. The Earth's oceans are the main reservoir of bacteriophages. In every milliliter of water in the ocean there are as many as one hundred million (10^8) bacteriophages. Another major reservoir are the guts of mammals, where each gram of fecal matter can also contain up to one hundred million bacteriophages.

Bacteriophages are classified into the 19 recognized bacteriophage families by the type of genome they possess and the shape of the protein coat particle, called the **virion**. All bacteriophages have one type of nucleic acid, which can be double-stranded DNA, single-stranded DNA, or RNA. Most bacteriophages that have been identified and characterized contain double-stranded DNA. Only two bacteriophage families have RNA genomes, which is why we tend to discuss DNA genomes of bacteriophages. The few bacteriophages with RNA genomes can contain either single-stranded or double-stranded RNA. Of the remaining 17 families, 15 have double-stranded DNA genomes and two have single-stranded DNA genomes. The virion morphology of bacteriophages is either polyhedral, often icosahedral, forming an isomeric shape, or spiral, forming a helical shape that is filamentous (**Figure 21.1**). The icosahedral bacteriophage structures sometimes have tails, with or without tail fibers, that are involved in attachment to bacterial cells and delivery of the bacteriophage nucleic acids into the bacterial cell.

Five bacteriophage families have envelopes around their protein coats, meaning that they are surrounded by a phospholipid membrane derived from the bacterial cell. In this way, these bacteriophages are similar to enveloped viruses that infect eukaryotic cells. There are proteins embedded within bacteriophage encapsulating membranes that are encoded by the bacteriophage genome. The bacterial membrane proteins are excluded from the region of the bacterial membrane that will be used to encapsulate the bacteriophages (**Figure 21.2**). This process is similar to the formation of animal cell targeting viruses and the deposition of viral proteins in their encapsulating membranes.

In addition to morphology, bacteriophages are characterized by being lytic or lysogenic. Lytic bacteriophages are actively virulent and will immediately replicate within the bacterial cell. Lysogenic bacteriophages can sometimes integrate their nucleic acids into the bacterial chromosome. The bacteriophage genome is therefore replicated within the bacterial genome each time the bacterial cell divides.

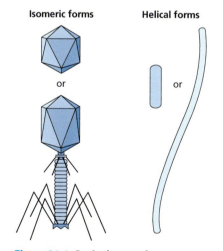

Isomeric forms   **Helical forms**

or   or

Figure 21.1 Basic shapes of bacteriophages. There are two basic morphologies of bacteriophages. The virion is either isomeric, in the shape of a polyhedral, such as an icosahedral, sometimes with a tail and tail fibers, or it is helical, in a spiral filamentous formation.

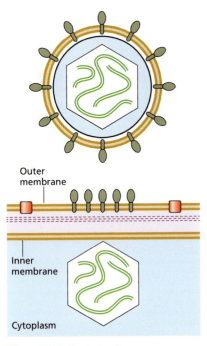

Figure 21.2 Bacteriophages with membrane envelopes. Some bacteriophages, such as φ6 shown here, are encapsulated by membrane envelopes, derived from the bacterial cell membrane. During formation (bottom panel), bacteriophage proteins (brown) are integrated into the bacterial membrane, excluding the bacterial surface proteins (red).

As discussed in Chapter 1, the bacteriophage genome integrated in the bacterial genome is called a prophage. Depending upon the nutritional status or stress experienced by the bacterial host cell, there may be a switch from lysogenic to the lytic phase, which then produces bacteriophage particles again.

Bacteriophages cannot replicate without bacterial cells

Bacteriophage replication requires bacterial cells because bacteriophages lack their own processes for protein synthesis. Replication of the bacteriophage requires it to adhere to the bacterial surface, penetrate the bacterial cell, replicate its nucleic acids within the bacterial cell, form the new bacteriophage particles, and then release the bacteriophage particles from the cell (**Figure 21.3**). The attachment and penetration are specific for the bacterial strain or species, dependent on surface structures of the bacteria.

Adsorption, adhesion of the bacteriophage to the bacterial cell, is the first stage required for bacteriophage replication to occur. In the process of adsorption, bacteriophage proteins bind to receptors on the bacterial cell surface, such as surface proteins including nutrient transporters and outer membrane proteins. In Gram-negative bacteria, some bacteriophages recognize LPS, whereas other bacteriophages recognize the teichoic acids of the Gram-positive cell wall. For bacteriophages with a head and tail structure, adsorption to specific bacterial surface receptors involves the tail fibers.

Penetration is the process of the bacteriophage nucleic acids entering the bacterial cell. This may be through rupturing the cell wall and injecting the nucleic acids into the cytoplasm or finding another route through the membranes and cell wall, such as through a bacterial porin protein. Early in the lytic cycle, bacteriophage proteins digest the bacterial genome. This shuts off production of bacterial replication and transcription, allowing the bacteriophage free reign over the bacterial systems needed for its own replication. In the **eclipse phase**, the new bacteriophages are made, including the nucleic acids and capsid, the protein coat of the bacteriophage particle. The mature bacteriophages are then formed and the bacterial cell is lysed to release them. Replication produces 100–200 new virions that are released upon lysis of the bacterial cell.

Figure 21.3 Replication of bacteriophages. Bacteriophages replicate within bacterial cells following attachment and injection of the bacteriophage genome into the cell. Once new bacteriophage particles are made, they are released from the bacterial cell via lysis.

Some bacteriophages enter latency for a period before replication

Although lytic bacteriophages had been studied for quite some time, it wasn't until Esther Lederberg described the behavior of lambda bacteriophages in 1950 that the capacity for a lysogenic phase in some bacteriophages was appreciated. A lysogenic bacteriophage enters latency, meaning that it does not replicate immediately, but becomes quiescent within the bacterial cell (**Figure 21.4**). The bacteriophage genome is still replicated, but the entire bacteriophage is not. The nucleic acid enters the cell and rather than triggering the digestion of the bacterial DNA, the bacteriophage DNA joins it, integrating into the bacterial chromosome. No bacteriophage capsid proteins are made. The bacteriophage genome within the bacterial genome has become a prophage and is replicated as the chromosome is replicated, being copied into each daughter cell. Bacteria containing prophage are referred to as **lysogens**. This latent state can be interrupted, triggering the initiation of replication of the bacteriophage and a lytic cycle. This can occur spontaneously or be induced by UV radiation including sunlight, by antibiotics, by alkylating agents, and by cellular stress.

There are also intermediate types of bacteriophage infection that are neither lytic nor lysogenic, although these are perhaps less common. For example, chronic infections result in bacteriophages budding from the bacterial cell, rather than lysis (**Figure 21.5**). Pseudolysogenic infections fail to properly integrate into the chromosome and either segregate out from one daughter cell at each bacterial division or enter lytic replication of bacteriophages (see Figure 21.5). In pseudolysogeny, the bacteriophage fails to either replicate in a lytic cycle or integrate as a prophage in a lysogenic cycle. Instead, the bacteriophage nucleic acids remain extrachromosomal. Ultimately, conditions within the bacterial cell enable the lytic or lysogenic cycles to occur, or the bacteriophage nucleic acids are lost.

Integration and excision of bacteriophage DNA into the bacterial chromosome relies on both bacterial and bacteriophage factors. Bacteriophage DNA, such as λ described in more detail later, circularizes upon entry into the cell. The *attP* site on the circular bacteriophage DNA and *attB* on the circular bacterial DNA recombine, catalyzed by the bacteriophage protein Int, an integrase, and bacterial protein IHF, integration host factor (see Chapter 5). This recombination process forms *attL* and

Figure 21.4 Lysogenic bacteriophages. Lysogenic bacteriophages integrate their nucleic acids into the bacterial chromosome.

Adsorption of bacteriophage

Penetration of phage DNA

Integration into the chromosome

Two daughter cells containing prophage

Bacterial genome replication and cell division

Figure 21.5 Chronic and pseudolysogenic bacteriophage infections. Chronic bacteriophage infections are similar to lytic infections, however, the bacterial cells do not lyse. Instead, the bacteriophages bud from the bacterial membranes and leave the bacteria intact. Pseudolysogenic infections do not integrate into the chromosome, therefore, the bacteriophage nucleic acids do not replicate with the bacterial DNA and do not segregate into daughter cells upon division, as they would in lysogens.

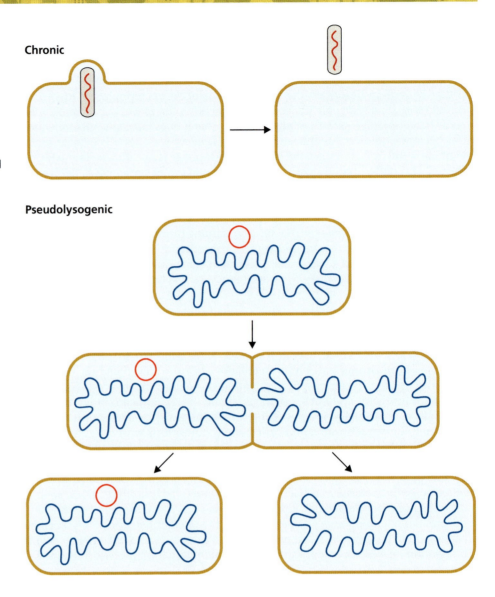

attR at either end of the integrated prophage (**Figure 21.6**). Upon induction of lysis, Int is again involved, along with Xis, a bacteriophage excision protein, which together remove the prophage and reform *attB* in the bacterial chromosome. This integration and excision system has been exploited in the development of cloning systems, such as Gateway, as discussed in Chapter 18.

The MS2 bacteriophage has a very small genome and was the first genome sequenced

Bacteriophage MS2 is an RNA bacteriophage and has the distinction of being the first complete genome to be sequenced, in 1976. This sequencing used a two-dimensional fractionation method that was only possible with RNA and pre-dates

Figure 21.6 Integration and excision of the bacteriophage genome. To generate the prophage, the bacteriophage genome enters the bacterial cell cytoplasm as linear DNA, which then circularizes. The circular bacteriophage genome recombines with the bacterial genome at the *attP* and *attB* sites, respectively. Integration requires the bacteriophage Int and bacterial IHF proteins. Excision needs the additional Xis bacteriophage protein and reforms *attP* and *attB* from the *attL* and *attR* sequences that flank the prophage.

Sanger sequencing. The genome is only 3,569 bases, encoding four proteins, three of which overlap.

The bacteriophage MS2 is 27 nm across, made of one copy of the maturation protein, encoded by the gene *mat*, and 180 copies of the coat protein, encoded by the gene *cp*. These proteins make an icosahedral shape capsid (**Figure 21.7**). The receptor for MS2 bacteriophage is the conjugation pilus encoded by the F plasmid.

Important discoveries about genetics have been made by studying bacteriophage λ

Bacteriophage λ was discovered in 1950 by Esther Lederberg and infects *Escherichia coli*. The virion structure has a capsid head, tail, and tail fibers (**Figure 21.8**). The head contains the double-stranded DNA, which is injected into the *E. coli* through the tail. There are approximately 1,000 proteins in one λ virion in total, with these being multiple copies of 12–14 different proteins.

The double-stranded DNA genome is 48,490 bases, plus 12 bases of single-stranded DNA at the 5' ends of each strand. These "sticky ends" of the λ genome are the *cos* site, which is responsible for the circularization of the linear genome once it is within the bacterial cytoplasm. The circularized double-stranded DNA within the bacterial cell is therefore 48,502 bp.

Bacteriophage λ uses the bacterial maltose porin for adsorption and through this its genome gains entry into the bacterial cell (**Figure 21.9**). After passing through the maltose porin, LamB, into the periplasm, the bacteriophage DNA passes through the mannose permease complex in the inner membrane to reach the cytoplasm, where it circularizes. It is a model for the study of lysogeny and from the study of λ a great deal has been learned about insertion and excision from the chromosome, where the *attB* site on the bacterial genome in *E. coli* sits between the *gal* and *bio* operons. Additionally, study of the expression of the λ genes has been insightful in the understanding of genetic regulation in general.

Figure 21.7 Bacteriophage MS2.
The structure, morphology, and size of bacteriophage MS2. The capsid is made of the maturation protein and 180 copies of the coat protein. The genome is 3,569 bases of RNA, encoding four proteins.

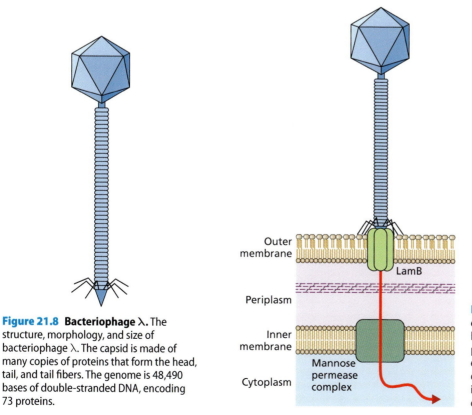

Figure 21.8 Bacteriophage λ. The structure, morphology, and size of bacteriophage λ. The capsid is made of many copies of proteins that form the head, tail, and tail fibers. The genome is 48,490 bases of double-stranded DNA, encoding 73 proteins.

Figure 21.9 Bacteriophage λ genome entry via LamB. The receptor for bacteriophage λ on *E. coli* is the maltose porin protein LamB. Binding to this porin enables the genome to pass through the outer membrane. It then passes through the inner membrane via the mannose permease complex to reach the bacterial cytoplasm.

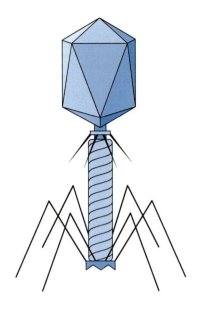

Figure 21.10 Bacteriophage T4. The structure and mechanism of nucleic acid injection into *E. coli* by bacteriophage T4. The tail of bacteriophage T4 has a contractile outer sheath and an inner core that is a rigid tube. When the tail fibers bind to the outer membrane, the sheath contracts and the tube pierces the outer membrane, peptidoglycan, and inner membrane, depositing the double-stranded DNA genome into the cytoplasm.

The T4 bacteriophage has a characteristic morphology

The bacteriophage T4 infects *E. coli* and was discovered in 1944 with six other T phages, T1–T7. The T4 genome is 168 kb, present as double-stranded DNA, which encodes 289 proteins.

Structurally, the T4 bacteriophage also has an icosahedral head with a hollow tail and tail fibers, like λ, and has a lytic lifecycle when infecting *E. coli*. The tail passes into the bacterial cell to infect and deliver the DNA (**Figure 21.10**). The tail of the T4 bacteriophage pierces the bacterial cell to inject the bacteriophage genome into the cytoplasm. The tail is made of a sheath that is able to contract, surrounding a rigid tube. At the bottom of the tail is a base plate, to which are connected the tail fibers that secure the bacteriophages to the outer membrane. When the sheath contracts, the inner tube of the tail is driven through the *E. coli* outer membrane, peptidoglycan cell wall, and inner membrane.

This tail resembles the bacterial Type 6 Secretion System (see Chapter 9) in structure and function (**Figure 21.11**). The contractile nature of the T4 bacteriophage tail matches the contractile function of the Type 6 Secretion System and the ability of this bacterial feature to stab competitor cells in its niche, including other bacterial cells, fungal cells, and eukaryotic cells.

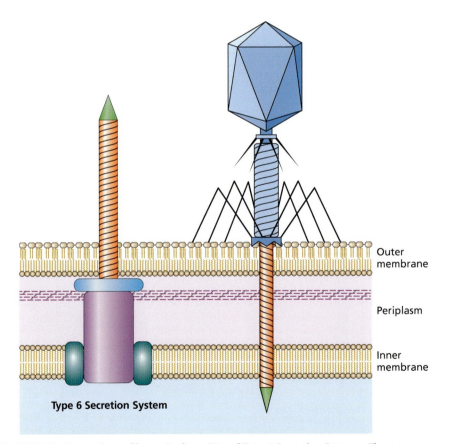

Outer membrane

Periplasm

Inner membrane

Type 6 Secretion System

Figure 21.11 Comparison of bacteriophage T4 and Type 6 Secretion Systems. The structure of the T4 bacteriophage is similar to λ and many other bacteriophage with a head and tail capsid structure. The tail of T4 is also similar to the bacterial Type 6 Secretion Systems, which evolved from bacteriophage structures.

The φX174 bacteriophage was the first DNA genome sequenced

The first complete DNA genome to be sequenced was from φX174, bacteriophages that infect *E. coli* and have a single-stranded DNA genome (**Figure 21.12**). This bacteriophage genome is 5,375 nucleotides and was the first sequence published with the Sanger method (see Chapter 11). The nature of the φX174 genome as circular DNA had previously been demonstrated by Walter Fiers and Robert Sinsheimer in 1962. The genome encodes 11 proteins, the genes for many of which overlap.

The adsorption target for bacteriophage φX174 is *E. coli* LPS (see Chapter 9). The capsid is an icosahedral shape without a tail. Bacteriophages with no tail rely on either bacterial structures such as pores or alteration of their own proteins to get their genomes into bacterial cells. For example, it was discovered in 2013 that the φX174 bacteriophage H proteins form together into a tube to deliver the bacteriophage DNA into the bacterial cell (**Figure 21.13**).

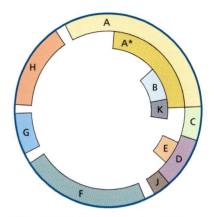

Figure 21.12 Bacteriophage φX174 genome. This is a representation of the genes encoded by the bacteriophage φX174 genome, which was the first DNA sequence to be determined using the Sanger sequencing method. This identified the coding regions of the bacteriophage.

Transduction is an important source of horizontal gene transfer for bacteria

Bacteriophage transduction is involved in bacterial evolution, as are transformation and conjugation. Prophage genes substantially increase the virulence potential of bacterial strains, contributing such genes as extracellular toxins, proteins involved in serum resistance, and in some cases antibiotic resistance genes. Not only do bacteriophages have the ability to integrate their own genetic material into the bacterial genome, but they can also mobilize other DNA. Some bacteriophages package the nucleic acids within the cell, whether it is bacteriophage in origin or bacterial.

Generalized transduction packages random pieces of DNA into bacteriophage capsids (**Figure 21.14**). This random DNA takes the place of bacteriophage genomic DNA and when infecting the next cell, it is the bacterial DNA that is transferred rather than the bacteriophage DNA. In **specialized transduction**, some bacterial DNA adjacent to the integrated prophage is excised with the prophage upon transition from lysogenic to lytic phase (see Figure 21.14). The bacterial DNA and bacteriophage DNA are taken as one piece into the bacteriophage capsid and released from the cell.

The prophages of group A streptococci are responsible for much of the genetic diversity of the species. Within the prophage genomes are genes that contribute to the pathogenicity of the various lineages of group A streptococci. Serotype M3 strains tend to cause deep tissue streptococcal infections, such as necrotizing fasciitis, as well as pharyngitis. Group A streptococcus serotype M3 strain MGAS315 has in its genome six different prophages containing six virulence factors. Prophage genes *speK*, *sla*, *sdn*, *speA*, *mf*4, and *ssa* encode streptococcal pyrogenic endotoxin K (SpeK), phospholipase A(2) (Sla), streptodornase (Sdn), streptococcal pyrogenic exotoxin A (SpeA), mitogenic factor 4 (Mf4), and streptococcal superantigen (Ssa).

Bacteriophage φX174 DNA

Bacterial cell

H proteins

Figure 21.13 The temporary tail of bacteriophage φX174. To deliver its single-stranded circular DNA genome into the *E. coli* cytoplasm, bacteriophage φX174 targets LPS on the surface of the bacterial cell, then polymerizes copies of the H protein into a tube through the outer membrane, peptidoglycan, and inner membrane. The circular genome of the bacteriophage is then able to pass through to the bacterial cytoplasm, at which point the H proteins dissociate.

Figure 21.14 Generalized and specialized transduction. Bacterial DNA can end up packaged within bacteriophages. In generalized transduction (left), random DNA is packaged into the capsid, which can include the bacterial DNA (blue), as well as the prophage genome (red). In specialized transduction (right), some bacterial DNA (blue) may be excised with the prophage (red) and end up packaged into the bacteriophages. In both cases, bacterial DNA is carried by the bacteriophage virion to the next bacterial cell.

These are induced in expression upon co-culture with human epithelial pharyngeal cells, exposure to hydrogen peroxide, or damage of DNA by mitomycin C. The various conditions *in vitro* mimic environmental signals that would occur *in vivo*, resulting in differential expression of prophage genes in the various host niches.

Bacteriophages also contribute to bacterial evolution through chromosomal rearrangements

Prophages are also involved in chromosomal rearrangements. Bacteriophage sequences can account for as much as 20% of a bacterial genome. Some of this sequence is intact bacteriophage genomic information capable of completing lytic replication, however, there are other bacteriophage sequences within bacterial genomes that are remnants of prophages. These may have been left behind in the genome by incomplete excision, may have been rendered inactive as bacteriophages by gene degradation, or may have been horizontally transferred into the chromosome or plasmid. The prevalence of bacteriophage sequences among the bacterial nucleic acids increases the likelihood that these sequences will mediate rearrangements by homologous recombination.

In *Neisseria gonorrhoeae*, for example, there are multiple prophages integrated into the chromosome, which contains regions that are homologous to each other. Comparative analysis between completely assembled circular genome sequences of *N. gonorrhoeae* strains revealed large segments of reordered and rearranged DNA sequence, some of which has been mediated by the presence of integrated bacteriophage sequences.

Changes in bacterial genomes by prophages can occur due to homologous recombination between similar prophage sequences integrated into the chromosome or mobilization of prophages from one area of the chromosome to another via excision and reintegration (**Figure 21.15**). This latter process may result in the translocation of bacterial genetic information as well as prophage sequence, just as excision may sometimes result in specialized transduction of bacterial sequences to new cells (see Figure 21.14).

Bacteriophage and prophage genome evolution can provide interesting insights

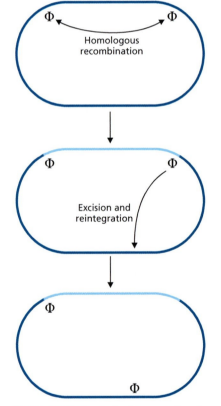

Figure 21.15 Bacteriophage-mediated chromosomal rearrangements. The presence of prophages within a bacterial chromosome or plasmid can result in rearrangements. This can occur due to homologous recombination between two prophage genomes, indicated here with Φ. This can also occur when the prophage excises and reintegrates elsewhere, sometimes taking some of the bacterial sequence with it in the process.

As many more bacterial genomes are entering the public sequence databases, so too are many prophage genomes. This provides opportunities for the study of bacteriophages via bioinformatics. One interesting investigation has explored the prophages within *Helicobacter pylori*. This bacterial species colonizes approximately half of all humans and the genotypes of this bacterial species can be linked to geographic origins of human migration across the globe. *H. pylori* and modern humans have evolved together with our ancestors carrying the bacteria with them in their stomachs as they migrated, passing the strains from mother to child. Like the bacterial genes, genes of the *H. pylori* resident prophages divide into populations with geographic distributions that correlate with human migration (see Chapter 17 and Figure 17.16).

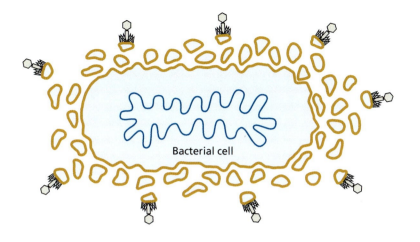

Figure 21.16 Outer membrane vesicles to survive bacteriophages. Blebs of the outer membrane have the same surface proteins and structures as the bacteria that they came from, therefore, bacteriophage adsorption occurs on outer membrane vesicles (OMVs) as readily as it does on bacterial cells. This can help protect the bacterial cell, by diverting the bacteriophages to the OMVs.

Bacteria have evolved strategies to avoid bacteriophage infection

To avoid infection by bacteriophages, bacteria can take several approaches. One is to avoid infection entirely by changing the host cell receptor on the surface of the bacteria. If bacteriophage adsorption is prevented, then so is infection. This can be accomplished through mutation of the gene encoding the receptor itself or concealing the receptor with a physical barrier such as a capsule or other modification. Phase variation (see Chapter 12) can also play a role, in which the receptor is only expressed some of the time, reducing risk to the bacterial cell.

Release of **outer membrane vesicles** (**OMVs**) can also provide a survival strategy. OMVs are blebs of bacterial membrane that are shed from the surface of the living bacterial cells (**Figure 21.16**). These contain proteins and other features characteristic of the surface of the bacteria. Bacteriophages can be sequestered away from the bacterial cell by adsorption to the OMVs instead of the bacterial cell itself. These decoys reduce infection of bacterial cells and increase their chances of survival.

Another strategy is to block entry of the bacteriophage nucleic acids into the bacterial cell. This uses a superinfection exclusion system, which is actually encoded by bacteriophages. Superinfection exclusion systems stop bacteria that have been infected by bacteriophages from becoming infected by new competitor bacteriophages. The resident bacteriophage produces a membrane-associated protein that interferes with the injection of new bacteriophage nucleic acids into the cell.

Even if bacteria become infected by bacteriophage nucleic acids, they can still fight back

Even if the nucleic acids of a bacteriophage make it into a bacterial cell, there are mechanisms whereby the bacteria can prevent the infection from destroying the bacteria. Restriction-modification systems are one means of protecting the bacterial cell from invading DNA, whatever the source. Therefore, when invading bacteriophage DNA enters the bacterial cytoplasm it is subject to restriction digestion by the resident bacterial enzymes. As described in Chapter 10, the bacterial DNA is protected from digestion by restriction-modification system methyltransferases, enzymes which modify specific recognition sites through methylation. The unmodified bacteriophage DNA is therefore not protected and is digested by the bacterial restriction enzymes before it can either replicate or integrate into the chromosome (**Figure 21.17**). Some bacteriophages are able to incorporate modified bases into their genomes as a mechanism of restriction-modification system resistance. To counter this bacteriophage genome modification, some bacterial restriction enzymes have recognition sites that include modified bases; although the sequence will be present in the bacterial genome, it will not be modified, therefore, the enzyme only digests foreign modified DNA.

Figure 21.17 Restriction digestion of bacteriophage DNA. When the bacteriophage DNA enters the cell, it is subject to the restriction-modification system. If the bacteriophage genome contains any recognition sequences for restriction enzymes expressed by the bacteria, the bacteriophage DNA will be digested. It is therefore neither replicated nor integrated into the chromosome.

Figure 21.18 Abortive infection system. Bacterial population level resistance to bacteriophages means that the infected cell sacrifices itself, but the surrounding cells survive. The bacteriophage replication is interrupted, resulting in premature lysis of the bacterial cell. Because no intact bacteriophage particles are made, the neighboring bacterial cells in the population survive. Compare to Figure 21.3.

Bacteria have also evolved a type of immune system against foreign DNA, the CRISPR/Cas systems that have been described previously (see Chapter 13). With this system, bacteria are able to build up a bank of foreign DNA sequences that can be recognized for elimination. This includes invading bacteriophage DNA. In addition to the CRISPR/Cas and restriction-modification systems that have been explored earlier in this book, there is another mechanism of resistance to bacteriophage infection that can act after entry of the nucleic acids into the cell.

As has been seen throughout this book, not all bacterial survival strategies involve survival of the individual bacterial cell, but rather take a population view. One means of surviving bacteriophage infection also takes this population level approach. An **abortive infection system** results in the death of the infected bacterial cell, however, the surrounding population is spared (**Figure 21.18**). The abortive infection system is genetically encoded on a mobile element such as a plasmid or even a prophage. The gene products involved disrupt bacteriophage replication so that mature, infectious bacteriophages are not successfully made. The disruption can happen at various stages of the process, but the end result is that replication does not occur and infection does not spread in the bacterial population.

Bacteriophage resistance that fights back and uses the bacteriophages for its own ends

The final mechanism of bacteriophage resistance to be discussed in this chapter is not so much resistance to bacteriophage infection, as a means of co-opting bacteriophages, allowing bacterial pathogenicity islands to hitchhike a ride to their next destination. This phenomenon has been well studied in *Staphylococcus aureus* where phage-inducible chromosomal islands, the *S. aureus* pathogenicity islands, are able to use incoming bacteriophages as a means of horizontally transferring to neighboring bacterial cells. These pathogenicity islands (see Chapter 6) have evolved mechanisms that enable the sequence of the island to be excised from the chromosome, replicated, and packaged into a bacteriophage capsid as a prophage would.

Upon infection with a bacteriophage, the *S. aureus* pathogenicity island encoded determinants limit the replication of the bacteriophage genome, in preference for its own excised sequence. The size of the capsid is also altered to the smaller capacity needed for its own DNA, further influencing the preference for mature particles with *S. aureus* pathogenicity island DNA, rather than bacteriophage DNA, packaged within.

Figure 21.19 Phage-inducible chromosomal islands. Some pathogenicity islands use infection by bacteriophages as a means to spread to new bacterial cells. Rather than succumbing to infection, the bacterial cell shuts down the replication of the bacteriophage genome, excises the pathogenicity island, replicates it, and packages it into the bacteriophage capsid. The pathogenicity island is then released with the lysis of the bacterial cell, spreading to other bacteria.

When the capsids are assembled with the *S. aureus* pathogenicity island DNA within them, these particles are released with the lysis of the bacterial cell. This process stops the spread of the original infecting bacteriophages. Its DNA has been excluded in preference for that of the pathogenicity island. Instead of the bacteriophages spreading to new bacterial cells, the pathogenicity island will spread to the surrounding cells (**Figure 21.19**). These particles themselves do not induce a lytic phase when they infect. Instead, when the DNA enters the new bacterial cell, the pathogenicity island will be potentially integrated into the bacterial chromosome.

Bacteriophage therapy is a potential alternative treatment for antimicrobial resistant bacteria

Phage therapy uses lytic bacteriophages to kill their target bacteria, without harming the host organism infected by the bacteria. With the advent of broad-spectrum antibiotics in the early to mid-1900s, research into bacteriophage therapy was discontinued in many areas of the world. Notably, therapeutic use of bacteriophages is already integrated into healthcare in Eastern Europe and nations of the former Soviet Union, where it has been in use for decades before the resurgence in interest elsewhere in recent years, due to multidrug antibiotic resistance. Species and strain specificity of bacteriophages makes them both ideal for treatment, as they will not indiscriminately eliminate other members of the microbiome like antibiotics, yet also more difficult to use. The specific etiological agent must be identified and its immunity to the bacteriophage treatment determined, which can be time consuming.

There are, however, potential benefits of the use of bacteriophages to target bacterial infections. There have been cases of success with phage therapy, saving lives and showing promise in clinical trials. Lytic bacteriophages have been demonstrated to disrupt *in vitro* biofilms of *Acintobacter baumannii*. In 2017, a patient in the US was successfully treated for *A. baumannii* pancreatic pseudocyst infection with bacteriophages administered intravenously. Phage therapy against *Pseudomonas aeruginosa* is promising, particularly to assist patients with cystic fibrosis suffering from chronic lung infections. Some of the bacteriophages considered for therapy target the bacterial efflux pumps, therefore, bacteriophage resistance would render the *P. aeruginosa* susceptible to antibiotics.

Further research into bacteriophages and bacteriophage therapy is needed. The restriction of bacteriophages to specific bacteria may not be universal; there are many uncharacterized bacteriophages and broad-spectrum therapeutic phages that may yet be isolated. Ecological and metagenomics studies suggest that although many of the well-studied bacteriophages have narrow specificities for bacterial hosts, there are others with broad host ranges.

Key points

- Bacteriophages are classified by their shape and by the type of genome they carry.

- Bacteriophages replicate by one of two means, either a lytic cycle that immediately makes more phage particles or a lysogenic phase, where the bacteriophage genome is integrated into the bacterial genome and replicates there.

- The first genome ever sequenced was of bacteriophage MS2, an RNA genome of 3.5 kb.

- The bacteriophage λ uses the surface porin for maltose to pass its DNA through the outer membrane of *E. coli*.

- The bacteriophage T4 has a tail with a contractile sheath around a rigid core that it uses to pierce the outer membrane, peptidoglycan, and inner membrane of *E. coli* to inject its DNA into the bacterial cell.

- Bacterial DNA can be horizontally transferred between cells by bacteriophages via transduction, either generalized transduction, where random DNA is packaged in the capsid, or specialized transduction, where some bacterial DNA goes with the bacteriophage genome.

- Prophages may account for as much as 20% of a bacterial genome and may be responsible for chromosomal rearrangements.

- Bacterial cells have evolved several strategies to avoid or survive infection by bacteriophages.

- Bacteriophages have found some ways to circumvent bacterial survival strategies evolved against them.

- Some bacterial pathogenicity islands have evolved to use invading bacteriophages as a mechanism to spread to other bacterial cells via transduction.

- Bacteriophage therapy has been used for decades in some parts of the world and is being investigated for more widespread use in the face of multidrug-resistant infections.

Terms, questions, and discussions

Key terms

Define the following terms introduced in this chapter. Check your answers using the definitions in the Glossary. These terms are also available as Flashcards on-line.

Abortive infection system	Generalized transduction	Specialized transduction
Adsorption	Lysogens	Virion
Eclipse phase	Penetration	

Self-study questions

Answer each question using 50–100 words or a table or labeled diagram. Advice on where to find answers to these questions is available on-line.

1. How many different recognized bacteriophage families are there and how are they classified into these families?

2. What characteristic do five of the bacteriophage families share with some eukaryotic viruses? How is this characteristic achieved?

3. Draw the lytic phase of bacteriophages. Include the key important stages.

4. Draw the lysogenic phase that is possible for some bacteriophages. Include the key important stages.

5. In 1976, the first genome was sequenced. What was it? What were some of the key features of this genome?

6. Describe how the bacteriophage λ DNA gets inside the *E. coli* cell and what happens to it once it is there.

7. Describe how the tail of the T4 bacteriophage delivers DNA into *E. coli*.

8. Compare generalized transduction and specialized transduction. Why would bacteriophages from generalized transduction not be able to produce bacteriophage particles when infecting the next bacterial cell? Why might some of the bacteriophages in specialized transduction also be deficient in production of new bacteriophages?

9. How can bacteriophages contribute to chromosomal rearrangements?

10. Name at least two ways bacterial cells can avoid bacteriophage infections. Describe how one of them is able to prevent bacteriophages from infecting the bacterial cell.

11. Name at least two defense systems bacterial cells have that can stop their destruction by bacteriophages once penetration has occurred.

12. Briefly explain how bacteriophage therapy selectively kills specific bacterial cells, but does not harm the host organism microbiome or host organism cells.

Discussion topics

These topics are presented for discussion in study groups, as part of class discussions, or on your own. These questions go beyond what is directly covered in this part of the book. Use the research literature and other reading to explore these topics in more depth. Tips to help prepare for topic discussions are available on-line.

13. In 1917, Félix D'Hérelle used isolated bacteriophages to control *Salmonella* in chickens. Decades later, bacteriophages were harnessed again for the same purpose to control *Salmonella* on organic chicken farms. Investigate this or another biotechnology application of bacteriophages in agriculture or another industry. Discuss the success of this intervention, the type of bacteriophages used, and the potential for expansion in the future.

14. Some bacteriophages are able to switch between the lytic phase and the lysogenic phase. Using your drawings from questions 3 and 4, indicate at what point bacteriophages capable of lysogeny, switch from the lysogenic phase in the drawing from 4 to the lytic phase in the drawing from 3. Likewise, at what stage does a bacteriophage change from lytic (3) to lysogenic (4)? Investigate and discuss what happens that triggers these switches.

15. Esther Lederberg discovered lambda (λ) bacteriophages and described lysogeny. She also made other contributions to microbiology and microbial techniques that contributed to bacterial genetics and genomics. The study of λ provided insight that was extrapolated across the field of genetics. Explore and discuss the contributions made by Esther Lederberg and at least one other scientist who made important contributions, but may not be well known, perhaps due to gender or race.

Further reading

Bacteriophages have been studied for just over 100 years

Ackermann H-W. Phage or phages. *Bacteriophage* 2011 Jan-Feb; 1(1): 52–3.

Some bacteriophages enter latency for a period before replication

Lederberg E. Lysogenicity in *Escherichia coli* strain K-12. *Microbial Genet Bull* 1950; 1: 5–8.

Antigenic variation Changes to antigens, generally bacterial surface structures, to prevent recognition by the host immune system, allowing for immune evasion

Antiparallel Describes the relative orientation of a string of molecules, where there are two strands parallel to one another, yet orientated in opposite directions

Antisense RNA A small ncRNA that is encoded on the opposite strand as the gene that it regulates, making it perfectly complementary to the mRNA

Antisense strand *See* negative strand

Anti-sigma factor A protein that binds to sigma factors, preventing it from binding to DNA

Anti-terminator A hairpin formation of mRNA that prevents the formation of a transcriptional terminator hairpin signal

Antivirulence agent A means to reduce the severity of an infectious disease by combating its virulence factors, such as secreted toxins

Aptamer RNA or DNA molecules that are able to form tertiary structures that are capable of binding to proteins or other targets

Archaea Organisms that are generally microscopic, yet are of a distinct evolutionary lineage from bacteria and eukaryotes

A-site *See* aminoacyl site

Assembly In genomics, the process of taking fragmented sequence data and piecing it together into longer contiguous sequences, based on overlapping sequence homology

ATP *See* adenosine triphosphate

Attenuation Reduction in production of complete transcripts

Attenuator A segment of mRNA sequence involved in attenuation of transcription

Autoinducer A molecule used in quorum sensing that is secreted by bacterial cells and sensed by bacterial cells to enable them to tally the presence of bacteria

Automated Sanger sequencing An enhancement of one of the first techniques for nucleotide sequencing, where use of fluorescently labeled nucleotides and computer-based detection automated much of the sequencing process

Autophosphorylate A protein that is able to add a phosphate, often from ATP, to itself

Autotransporter A type 5 secretion system protein that is able to transport itself out of the cell using a transporter domain that forms an outer membrane channel

Auxotroph A bacterial strain that has an auxotrophy

Auxotrophy Due to a genetic mutation, a bacterial cell requires a specific nutrient in the media to grow

Backbone In genetics, this refers to the sugar phosphate portion of a nucleic acid strand

Bacteria Organisms that are generally microscopic, single cells, of a distinct evolutionary lineage from eukaryotes and archaea

Bactericidal An antimicrobial that kills the bacterial cell

Bacteriocin Antimicrobial peptides produced by bacteria to eliminate competitor bacteria in a niche

Bacteriophage A small virus-like particle containing its own genetic material and capable of infecting bacteria

Bacteriostatic An antimicrobial that stops the growth of the bacterial cell

β barrel *See* beta barrel

Basal body *See* base complex

Base calling Identification of the individual nucleic acid sequence data from interpretation of the sequencing machine information; for example, this may be incorporation of bases in sequencing by synthesis and generation of light or release of hydrogen ions, or it may be differential through a nanopore

Base complex A multi-protein component of a type 3 secretion system that is embedded in the inner and outer membranes, spanning the periplasm

Basic local alignment search tool A computational method for finding sequences in a database similar to a query sequence

Beta barrel A tertiary amino acid structure formed by the stacking of beta sheets into a cylindrical structure

Beta sheet A secondary amino acid structure formed by hydrogen bonding between amino acids such that a sheet is made

Bioaugmentation Restoration of environmental sites contaminated with pollutants through the addition of microorganisms specifically chosen to address particular issues in that environment, with the goal of accelerating the bioremediation process

Bioenergy Generation of energy from biological sources, ideally renewable

Biofilm A community of microorganisms that together form a structure that is more resilient than the cells are individually

Bioinformaticians People who use computers to analyze biological data, bioinformatics

Bioinformatics Computational and software-based approaches to understanding biological data, such as DNA sequence data

Bioreactor A vessel for the long-term growth of microorganisms in a controlled environment

Bioremediation Restoration of environmental sites contaminated with pollutants through the introduction or assistance of microorganisms

Biostimulation Restoration of environmental sites contaminated with pollutants through the addition of nutrients and other factors that aid microorganisms naturally occurring in the environment and that are already conducting bioremediation, with the goal of accelerating these processes

Biotechnology The use of biological organisms for the improvement or betterment of human life

Biotreatment Removal of pollutants and contamination from waste water and other by-products of industry by microorganisms before they enter the environment

BLAST *See* basic local alignment search tool

BLAST hits The results from a BLAST search; sequences in the database that match the query sequence

BLASTN *See also* BLAST, a specialized form of BLAST for comparing a nucleotide sequence with a nucleotide sequence database

BLASTP *See* also BLAST, a specialized form of BLAST for comparing an amino acid sequence with an amino acid sequence database

BLASTX *See* also BLAST, a specialized form of BLAST for comparing a nucleotide sequence with an amino acid sequence database by first translating the nucleotide sequence into its six frame protein translation

Blunt ends Ends of a DNA double helix generated by a restriction enzyme, where the DNA is cut straight across so that there are no overhanging nucleotides

Branched chain amino acids Amino acids with branched side chains

Broad-spectrum antibiotic An antimicrobial agent that acts on a wide variety of bacterial species

cAMP *See* cyclic AMP

cAMP receptor protein A protein, also known as CAP, that is able to bind to another molecule within the cell, cAMP, and is then able to bind to DNA, thus exerting repression of gene expression

CAP *See* catabolite activator protein

Capsule An extracellular polysaccharide coating on the surface of some bacterial species

Carboxyl group A carbon, two oxygens, and a hydrogen, found at one end of an amino acid

Cas *See* CRISPR associated genes

Catabolite activator protein A protein that is able to bind to another molecule within the cell and able to bind to DNA, thus exerting catabolite repression

Catabolite repression The reduction or prevention of transcription due to the presence of molecules within the cell

CDS *See* coding sequence

Cephalosporins A class of antibiotics that contain the characteristic β-lactam ring structure, as well as an altered adjacent ring structure that renders this class resistant to standard β-lactamases

Chaperone A protein that assists in the folding or stabilization of another protein

Chaperones Proteins that aid in the folding of other proteins

Chaperone-usher secretion A type 5 secretion system where three proteins work to make an outer membrane channel, properly fold a secreted protein, and export it

Cheater A bacterial phenotype that is able to continue in a hostile environment due to the presence of other organisms in the niche with a favorable phenotype that neutralize the hazards of the environment

ChIP-Seq The sequencing of chromatin immunoprecipitation to reveal the locations across the genome where proteins bind to the DNA

Chromatin Proteins that interact with DNA structurally, often causing it to become compact

Chromatin immunoprecipitation The identification of DNA bound by protein through use of antibodies directed against the protein to separate out the bound DNA

Chromatin immunoprecipitation sequencing *See* ChIP-Seq

Chromosome Discrete DNA that contains essential genetic material and is copied by replication; one or more chromosomes make up a genome

Cis Referring to chemical structures those being on the same side as one another; referring to genetics those being close

Clonal complexes Phylogenetic contexts for bacterial isolates as defined by sequence typing methods such as MLST

Clone In molecular genetics, this refers to bacterial cells that are in some way altered in the laboratory

Closed complex An assembly of RNA polymerase and DNA at a promoter region where hydrogen bonds between bases of opposite DNA strands remain intact

Clustered regularly interspersed short palindromic repeats A genetic system that enables bacteria to recognize foreign nucleic acid sequences such as bacteriophage

Coding region *See* coding sequence

Coding sequence An open reading frame that has an appropriately located initation codon such that it could encode a gene

Codon A set of three nucleotides that corresponds to a sequence on a particular tRNA carrying a particular amino acid, thus the three letters encode a specific component of a protein

Codon table A listing of which three bases correspond to which amino acids

Cold-shock proteins Proteins that protect the bacterial cell from the stresses experienced due to decrease in temperature

Colony PCR Use of a single bacterial colony, or DNA from one, as the template in PCR, *see* also PCR

Comparative genomics Analysis of genome sequence data when two or more genome sequences are evaluated against one another to determine similarities and differences between them

Comparative modeling *See* homology modeling

Compatible cohesive ends Reference to the generation of restriction enzyme cut end overhangs that contain the same bases, even though they were generated by different enzymes

α-Complementation To indicate the successful cloning of an insert, this engineered system will generate blue clonies with no insert, white with a cloned insert

Complementation In molecular genetics, this adds back a gene that has been disrupted in a knockout mutant to check the phenotype is related to that gene and not another change in the genome

Concatemerization Generation of long, continuous DNA molecules that contain repeated or multiple copies of the same sequence

Conjugation The transfer of DNA from one bacterial cell to another by virtue of the F plasmid and conjugation pilus

Conjugation pili A specialist appendage of a bacterial cell that is encoded by the F plasmid and which is a conduit for transfer of DNA from donor cell to recipient cell

Consensus sequence A representative sequence arrived at based on the sequences that are generally present across a number of strains and/or species

Conserved domain A functional portion of a protein that is similar between proteins of similar functions, even if the whole of the protein does not share similarity, *see* domain

Conserved hypothetical protein A hypothetical protein that has homology with other hypothetical proteins in other bacterial species

Constitutive expression Expression of a transcript or protein at a relatively uniform level in the cell

Construct DNA that has been generated in the laboratory, generally through joining together different sequences

Contig A stretch of contiguous sequence data, often assembled from separate sequencing reactions that generated data with overlapping identical sequences

Convergent evolution Genetic sequence changes observed in separate bacterial genomes where the similarity in selective pressures results in similarity in gene changes

Core genes A gene or genes that are common to all genomes of a particular species

Core genome All of the genes that are common to all genomes of a particular species

Corepressor A small molecule that binds to a repressor protein and activates it so that it can bind to DNA

Coupled transcription–translation Where the process of transcription (making mRNA) and translation (making protein) happen concurrently for a given gene

CRISPR *See* clustered regularly interspersed short palindromic repeats

CRISPR associated genes Genes associated with the CRISPR region of the genome, encoding proteins involved in the bacterial immune system

CRP *See* cAMP receptor protein

Cyclic AMP A form of AMP, adenosine monophosphate, that is present in the cell inversely to the presence of glucose

Cyclic di-GMP A secondary signal for the state of the bacterial cell made from two GTPs

Cytosine A nitrogenous base made of four carbons, three nitrogens, one oxygen, and five hydrogens, found in DNA and RNA

Data depth This is a measure of how many times each individual nucleotide has been sequenced in all of the separate sequencing reactions

Deamination Through the addition of water, an amine group is lost, generally from cytosine resulting in uracil

Deformylase An enzyme that removes a formyl group (−COH) from a protein; the first step in removing fMet from the start of a peptide

Deoxyribonucleic acid (aka DNA) the genetic material of bacteria that stably encodes their traits, comprised of deoxyribose sugars, phosphates, and the nitrogenous bases adenine, cytosine, guanine, and thymine

Deoxyribonucleoside triphosphate (aka dNTP) a deoxyribose sugar, three phosphates, and a nitrogenous base

Deoxyribose A sugar containing five carbons, four oxygens, and 10 hydrogens that forms the backbone of DNA, together with phosphate

Diguanylate cyclases Enzymes that forms cyclic di-GMP from two GTPs

Dihydrouridine A modified nucleoside present in some tRNAs and rRNAs

Dimer Two of the same protein that complex with one another

Dinucleotide Two nucleotides joined by a phosphodiester bond between the sugar of one and the phosphate of another

Directional cloning Use of restriction enzymes so that an insert will only be ligated into a plasmid in one specific orientation

Disulfide bond A bond formed between two sulfides, often within proteins and formed between two cysteine amino acids

Divergent promoters When two nearby promoters are transcribed in opposite directions and away from each other, one on one strand and one on the other

D-loop The loop from the first of three stem loop structures in tRNA, starting from the 5′ end

DNA *See* deoxyribonucleic acid

DNA gyrase A replication enzyme that unwinds the DNA double helix at the replication fork

DNA polymerase An enzyme that is able to form a new strand of DNA based on a single strand of DNA as the template

dNTP *See* deoxyribonucleoside triphosphate

Domain A functional part of a protein

Dosage effect As replication copies DNA, some genes are present in multiple copies, which impacts upon their expression

Double digest Use of two different restriction enzymes with different recognition sites simultaneously

Downstream After a gene or other genetic feature, 3′ of the feature

D-stem The stem from the first of three stem loop structures in tRNA, starting from the 5′ end

E value Short for expect value, this is the number of alignments a BLAST query would generate in a search of the database by chance

Eclipse phase In bacteriophage replication, when bacteriophage particles and the genome are replicated and the bacterial systems are shut down

Effector A protein that is exported from the cell by a type 3 secretion system

Effector molecule A non-protein component, often a metabolite, that is able to influence transcription

Efflux pump system Multiprotein systems that export harmful substances out of the bacterial cell, including antibiotics

Elongation (transcription) In genetics, the phase of the process of transcription whereby RNA is made from a DNA template through the addition of nucleotide bases and the formation of phosphodiester bonds along the backbone

Elongation (translation) In genetics, the phase of the process of translation whereby a polypeptide chain is made based on an mRNA sequence

Endonucleolytic Action upon a nucleic acid within the strand

Enhancers Features within a sequence that increase transcription

Epidemiology The study of the distribution of diseases, geographically and over time

E-site *See* exit site

ESKAPE pathogens *Enterococcus faecium, Staphylococcus aureus, Klebsiella pneumoniae, Acinetobacter baumannii, Pseudomonas aeruginosa,* and *Enterobacter* species that are of clinical importance as pathogens

Essential gene A coding sequence required for life

Etiological agent The cause of a disease; the bacteria that is the source of the infection

Etiological agent That which is causing the disease, which can be a microorganism or toxin or other source

Eukaryote Also referred to as higher organisms, this form of life can be multicellular, with complex cell structures, of a distinct evolutionary lineage from bacteria and archaea

Evidence tag In annotation of sequence data, the scientific justification of the assigned annotation

Evolvability The capacity of the bacteria to evolve solutions to adapt

Evolve and resequence Growth of a bacterial culture for a period of time in a specific condition followed by sequencing of the genome again and comparison with the original data

Exit site A place within the ribosome where tRNA that has released the growing polypeptide chain goes before exiting the ribosome

Exonucleolytic Action upon a nucleic acid from the end of the strand

Exopolysaccharide Sugars secreted by bacteria into the surrounding environment that are a feature of biofilms

Expression platform The regulatory region of a riboswitch, which together with the aptomer region, can regulate expression of the mRNA

Expression profiling Assessment of the levels of transcription of genes

Extended −10 A conserved sequence sometimes present approximately 10 bases 5′ of the transcriptional start site that has high affinity for the sigma subunit of RNA polymerase, therefore no −35 region is needed

Extended spectrum β-lactamases A class of enzymes that have overcome the difference in the structure of cephalosporins that render them resistant to standard β-lactamases

F plasmid A plasmid containing the genetic features needed to make a bacterial cell a conjugation donor

Factor for inversion stimulation A nucleoid protein that contributes to bending of the DNA

Facultative aerobes Bacteria that are able to switch their metabolism and continue to grow in the absence of oxygen

Facultative intracellular Bacteria that can live extracellularly or can live within cells

FASTA A genetics file format, which includes a first line with information about the sequence, followed by the nucleic acid sequence

Feedback regulation The expression of the gene encoding a regulator, which is regulated by the regulator itself

Fertility factor The genetic features needed to make a bacterial cell a conjugation donor

F-factor *See* fertility factor

Fimbriae Thin, numerous bacterial surface structures made of protein

Fis *See* factor for inversion stimulation

Flagella Long thick bacterial surface structures made of protein that are capable of rotating to cause cell motility

fMet *See* N-formylmethionine

Fold recognition *See* protein threading

G+C The number of guanines plus the number of cytosines, an indicator calculation that can be performed on a sequence of DNA and give a figure characteristic of the DNA

Galactose A sugar generated from the breakdown of lactose by β-galactosidase

β-galactosidase An enzyme that breaks down lactose into its component sugars; glucose and galactose

β-galactoside permease A transport protein that transports lactose into the bacterial cell

GC skew A calculation that can be performed on a DNA sequence, G-C/G+C, which can help identify the Origin of Replication

Gene Genetic feature encoding a protein that is characterized as having an initiation codon, a termination codon, and for which a protein product has been demonstrated

Gene conversion A mechanism used in antigenic variation where the sequence of an expressed gene locus is changed through recombination with sequences at silent loci

Generalized transduction Due to random packaging of nucleic acids into the bacteriophage head, some bacterial DNA may be packaged and therefore carried to new bacterial cells

Genetic map Determination of the relative location of genes on the chromosome, which can be created using timed conjugation experiments

Genome The genetic material, containing all of the traits typically found in an organism

Genotype The genetic traits possessed within an organism's genome

GlcNAc *See* N-acetyl glucosamine

Global alignment Matching the similarity of sequences so that similar nucleotides or amino acids align with one another over the whole of the sequence; an alignment will be generated even if there is no similarity

Global regulator A protein that can alter transcription of several genes

Glucose A sugar essential for energy in bacterial cells

Glycoprotein A protein that has a sugar that has been added to it

Glycosylation The addition of a sugar to a protein

Group I introns Regions that are capable of splicing themselves out of a nucleic acid

Guanine A nitrogenous base made of five carbons, five nitrogens, one oxygen, and five hydrogens, found in DNA and RNA

Heat inactivated Some enzymes can be rendered inactive, unable to perform their enzymatic activity, after exposure to certain temperatures

Heat-shock proteins Proteins that protect the bacterial cell from the stresses experienced due to increase in temperature

Heat-unstable protein A nucleoid protein that contributes to bending of the DNA

α **helix** *See* alpha helix

Heuristic A computer science term for finding an approximate solution quickly

Hfr *See* high frequency of recombination

HGT *See* horizontal gene transfer

High frequency of recombination A designation for bacterial cells that have integrated the F plasmid into their genome

High-performance liquid chromatography A technique used to separate parts of a complex mixture for the purposes of identification and quantification

Histone-like nucleoid structuring protein A protein, which together with more of itself, binds to DNA and contributes to the compact nature of the nucleoid

H-NS *See* histone-like nucleoid structuring protein

Holoenzyme The whole of an enzyme with all of its subunits

Homodimer Two identical proteins that come together to form a quaternary structure, often seen in DNA binding proteins

Homology A measure, usually in percentage, of the identity or similarity of two DNA or protein sequences

Homology modeling Generation of a three-dimensional model of a protein based on experimental evidence of the structure of a homologous protein and the amino acid sequence of the protein of interest

Horizontal gene transfer The process of movement of DNA from one bacterial cell to another that then results in the incorporation of the DNA into the new cell

HPLC *See* high-performance liquid chromatography

HU *See* heat-unstable protein

Hydrogen bond A molecular connection between molecules formed by charge associated with hydrogen

Hydrolysis Through reaction with water, a chemical structure is split

Hydroxylation The addition of an oxygen (OH) to a protein

Hypothetical protein A predicted protein sequence from a CDS that is proposed to encode a protein but for which there is no experimental evidence or sequence homology suggesting a function

Ident value Short for identity value, this is the identity score between the aligned BLAST query and target sequences

Identity (amino acid) A measure, usually expressed in percentage, of how much two protein sequences match

Identity (DNA) A measure, usually expressed in percentage, of how much two DNA sequences match

IHF *See* integration host factor

In vitro A non-English term (note the italics) meaning in an artificial situation in a laboratory

In vivo A non-English term (note the italics) meaning in a natural situation within a living host or host cells

Indel A short addition, insertion, of one or a few bases or a short loss, deletion, of one or a few bases from DNA

Initial Darwinian Ancestor A hypothetical first organism on Earth, capable of passing on its genetic material

Initiation (transcription) In genetics, the beginning of the process of transcription, where transcription enzymes are recruited to promoters

Initiation (translation) In genetics, the beginning of the process of translation, where translation enzymes, mRNA, tRNA, and ribosome are recruited to begin making a protein

Initiation factor protein One of a set of proteins inolved in the initiation of translation

Inner membrane platform A protein component of a type 2 secretion system that is embedded in the inner membrane

Inosine A modified nucleoside present in some tRNAs

Integration host factor A nucleoid protein that contributes to bending of the DNA

Internalins Proteins that help bacteria to be internalized by host cells

Intracellular Bacteria that can live within other cells

Intrinsic transcriptional termination When the signal to stop transcription is a built in part of the sequence being transcribed

Introns Regions of a nucleic acid that are not present in the mature form of the DNA or RNA

Inversion When a segment of DNA has been flipped from one strand to another

IS elements Features within bacterial DNA that are capable of self-mobilization from one part of the DNA to another

Isoelectric point The pH at which a protein has no net electrical charge

Isogenic Two or more bacterial strains that have nearly identical genetic backgrounds, but which often differ by one or a discrete number of mutations

kb *See* kilobase

Kilobase One thousand bases of nucleic acid

Knockout In molecular genetics, this is a mutation that gets rid of the functional sequence of a gene through deletion and/or disruption of the gene sequence

β**-lactam** A class of antibiotics containing a characteristic ring structure of three carbons, one with a double-bonded oxygen, and a nitrogen

β**-lactamases** A class of enzymes that inactivate the class of β-lactam antibiotics by cleaving the β-lactam ring between the nitrogen and the carbon with adjacent double-bonded oxygen

Lactose A desaccharide sugar made up of glucose and galactose

Lagging strand During replication, this is the strand of the DNA that must be copied in short segments that proceed away from the replication fork

LAMP *See* loop-mediated isothermal amplification

Landmark database A small, highly curated database of sequences representing 27 genomes, including 11 bacterial species

LC/MS/MS *See* liquid chromatography tandem mass spectrometry

Leader A segment of mRNA sequence involved in attenuation of transcription

Leading strand During replication, this is the strand of the DNA that can be copied directly, proceeding toward the replication fork

Library In molecular genetics, this is a collection of bacterial strains that are identical apart from a single discrete mutation

Ligase An enzyme that joins together the ends of a nucleic acid backbone; used in replication and molecular biology

Lineage Bacterial cells that originated from the same precursor cell and thus have a common ancestor

Lipid A The lipid portion of LPS and LOS, which has a hydrophobic region that is embedded in the outer membrane

Lipidation The addition of a lipid to a protein

Lipooligosaccharide Molecular structures that are on the surface of a few Gram-negative bacteria that form the outer membrane with phospholipids

Lipopolysaccharide Molecular structures that are on the surface of most Gram-negative bacteria that form the outer membrane with phospholipids

Lipoprotein A protein that has had a lipid component added to it

Liquid chromatography tandem mass spectrometry A technique combining separation by HPLC with mass analysis by MS; sensitivity and specificity is increased by using two mass spectrometers in tandem

Local alignment Matching the similarity of sequences so that similar nucleotides or amino acids align with one another in the segments of the sequence in which they are similar; if there is no similarity, no alignment is generated

Locus identification number Sequential numbering of features in a genome sequence with unique identifiers

Loop-mediated isothermal amplification A technique used to amplify DNA at a constant temperature using a series of nested primers that generate self-priming loops

LOS *See* lipooligosaccharide

Low complexity regions Parts of a query sequence that will confuse a BLAST search due to the frequency of occurrence in databases, such as repetitive sequences

LPS *See* lipopolysaccharide

LPS core The portion of LPS and LOS that is directly attached to the lipid A; in LPS the O-antigen is attached to the LPS core

Lysine acetyltransferase An enzyme that adds an acetyl group to a lysine residue within a protein

Lysine deacetylase An enzyme that removes an acetyl group from an acetylated lysine within a protein

Lysogenic phase Describes when a bacteriophage goes dormant within a bacteria, with its genome integrated into the bacterial chromosome as a prophage

Lysogens Bacterial cells containing one or more prophage genomes integrated into their chromosome(s)

Lysosome A eukaryotic organelle containing lysosomal enzymes that are capable of digesting bacterial cells

Lytic phase Describes when a bacteriophage replicates its genome and constructs new phage particles that break free and are capable of infecting other bacterial cells

MALDI-TOF-MS *See* matrix-assisted laser desorption ionization-time-of-flight mass spectrometry

Mapping In the context of bioinformatics, alignment of sequence data to a reference sequence

Mass spectra The results produced by a mass spectrometer based on characteristics of the sample analyzed

Mass spectrometry Technology for the analysis of samples based on the mass of molecules within the sample

MATE family A class of efflux pump that uses sodium import into the bacterial cell to fuel export out of the cell

Matrix-assisted laser desorption ionization-time-of-flight mass spectrometry A technique used to identify features of a sample, separating aspects of the sample to produce a characteristic mass spectra

Max score Short for maximum score, this is a value assigned to a BLAST result based on aligned and matched sequences, taking account of those that mismatch and any gaps

Mb *See* megabase

MCS *See* multiple cloning site

Megabase One million bases of nucleic acid

Megaplasmid A very large plasmid

Membrane fusion protein A protein in the periplasm that bridges an inner membrane protein and an outer membrane protein, present in type 1 secretion systems

Messenger RNA Ribonucleic acids that have been copied from DNA in a process called transcription that carry the genetic make-up of a gene or genes

Metagenomic sequencing The sequencing of DNA from a sample taken from a natural environment, without discrimination or selection for organisms

Methionine aminopeptidase An enzyme that removes the N-terminal methionine from a peptide chain after the action of deformylase

Methylation The addition of a methyl group

Methylation The addition of a methyl group ($-CH_3$) to a protein

MFS family A class of efflux pump made of protein with several transmembrane helices

Microbiome The microorganisms present in a particular environmental niche, collectively

Microcolonies Small clusters of bacterial cells replicating in a localized area

MLD *See* mRNA-like domain

MLST *See* multi-locus sequence typing

Molecular biology The investigation of nucleic acids and/or proteins, including their structure, interaction, and regulation

Molecular cloning The generation of recombinant DNA using molecular biology techniques

Molecular mass *See* molecular weight

Molecular weight The mass of a protein or other molecule, based on the atomic weight of its parts

mRNA *See* messenger RNA

mRNA-like domain Part of a tmRNA that is like mRNA and carries a short sequence including a C-terminal peptide tag and a termination codon

MS *See* mass spectrometry

Multi-FASTA A genetics file format, which includes the information from more than one FASTA file, where there is more than one line with information about the sequence, followed by the nucleic acid sequence described

Multi-locus sequence typing A means of typing bacterial isolates that sequences a discrete set of genes and compares the data as a typing system

Multiple cloning site Engineered sites within plasmids that contain many, overlapping restriction enzyme recognition sites

Multiple sequence alignment Matching the similarity of three or more nucleotide or amino acid sequences along their length

MurNAc *See* N-acetyl muramic acid

Mutation A change in a DNA sequence

N-acetyl glucosamine One of two repeating units that make up the peptidoglycan of the bacterial cell wall

N-acetyl muramic acid One of two repeating units that make up the peptidoglycan of the bacterial cell wall

N-acyl homoserine lactones Signal molecules produced by bacteria for quorum sensing

NAG *See* N-acetyl glucosamine

NAM *See* N-acetyl muramic acid

Nanopore An artificially created pore that is on the scale of nanometers or smaller

Narrow-spectrum antibiotic An antimicrobial agent that acts on a single species of bacteria or a very few species

ncRNA *See* noncoding RNA

Needle A multi-protein component of a type 3 secretion system that extends through the base complex out of the bacterial cell, through which effectors pass

Negative strand The strand of DNA complementary to that encoding a protein

Next-generation sequencing A collective term for advances in sequencing technology that allowed high-throughput generation of nucleotide sequence data rapidly, inexpensively, and easily

N-formylmethionine At the start of production of any protein by the ribosome, this is the first amino acid

Nitrogenous base *See* nucleoside base

Noncoding RNA A small RNA that is transcribed, but not translated, and may be involved in regulation of gene expression

Non-stop decay Degradation of an mRNA by RNases as a result of a stalled ribosome that has not encountered a signal to terminate translation

Non-synonymous A change in a DNA sequence that results in a change in amino acid when translated

Nosocomial infections Hospital-acquired infections

N-terminal acetyltransferase An enzyme that adds an acetyl group to the N-terminal end of a protein

Nuclease Enzymes that degrade nucleic acids

Nucleic acid Individual components that make up DNA or RNA, comprising a sugar, a phosphate, and a nitrogenous base

Nucleoid The condensed mass of chromosomal DNA within a bacterial cell that can be seen using a microscope

Nucleoside base An adenine, cytosine, guanine, thymine, or uracil that is not connected to a sugar phosphate group

Nucleotide base Adenine, cytosine, guanine, thymine, or uracil with a sugar phosphate

Nucleotidyl transferase An enzyme that adds the CCA motif to the 3′ end of a tRNA forming the amino acid acceptor motif

O-antigen The portion of LPS that is farthest away from the surface of the cell and is the most variable in length and structure; not present in LOS

Obligate aerobes Bacteria that require oxygen to grow

Obligate intracellular Bacteria that can only live within other cells

Okazaki fragment The name given to the short segments of DNA made on the lagging strand in replication

Oligonucleotide A short string of nucleotides joined by phosphodiester bonds

OMVs *See* outer membrane vesicles

Open complex An assembly of RNA polymerase and DNA at a promoter region where hydrogen bonds have been broken and the DNA strands have separated

Open reading frame A region between two termination codons

Operator A term for the region 5′ of a gene that can be bound by a regulatory protein that influences the level of transcription

Operon Two or more co-transcribed and co-regulated genes

Opsonization Immunological targeting of a bacteria for phagocytosis by covering it with antibodies or complement

ORF *See* open reading frame

Origin of Replication A location within a DNA chromosome where replication consistently begins

Orthologous regions Regions of two different genomes that share homology; not necessarily defined by the presence of coding sequences, *see also* orthologue

Orthologue A homologous gene sequence between two bacterial species, where the two genes have a common ancestral origin, suggesting that the genes may have a common function

Outer membrane complex A protein component of a type 2 secretion system that is embedded in the outer membrane

Outer membrane vesicles In some bacterial species, portions of the outer bacterial membrane separate off from the bacterial cell as extracellular vesicles, also called blebs

Output domain The section of a response regulator protein in a two-component regulatory system that is activated through phosphorylation of the receiver domain, activating the regulatory function of the response regulator

Overexpression mutant A bacterial cell with a mutation that causes the gene to be expressed at higher levels than normal

Overlapping PCR Use of several rounds of PCR, with overlapping primer design, to generate a novel DNA sequence different from the original template, *see* also PCR

Oxidative folding The formation of disulfide bonds between cysteines in a protein

Pairwise alignment Matching the similarity of two nucleotide or amino acid sequences along their length

Palindromic A word, number, or sequence that can be read the same forward and backward, such as the word madam or the number 1234321

Pan-genome All of the genes that have been found within all of the genetic sequences from a particular species

Pan-genome The collective genetic features between bacterial genome sequences, including both the core genes and all of the accessory genes present in the set of bacterial genomes

Paralogue A homologous gene sequence within a bacterial genome, where the two genes have a common origin, having arisen through a gene duplication event, but which may now have diverged in function

Pathogenicity island A region containing genes involved in virulence or pathogenicity, which may have been horizontally transferred

PCR *See* polymerase chain reaction

Penetration Insertion, in the context of bacteriophage, the bacteriophage genome entering the bacterial cell

Peptide bond The bond formed between two amino acids, with the carboxyl group of one amino acid joining to the amino group of another

Peptide chain A series of amino acids connected by peptide bonds

Peptidyl site A place within the ribosome where tRNA is transferred from the A-site; this is where the growing polypeptide chain remains attached to a tRNA

Persister A bacterial phenotype that is able to continue in a hostile environment due to its slow growth

Phagocytosis The ingestion of cells or particles by other cells, often immune cells engulfing bacterial cells

Phagosome A eukaryotic organelle that has engulfed a cell or particle

Phase variation Stochastic switching of gene expression, generally ON to OFF to ON, but also in levels of expression

Phasevarion A set of genes that are regulated by a single phase variable gene, usually a modification component of a restriction-modification system

Phenotype The genetic traits being expressed by an organism at a given time

Phosphodiesterases Enzymes that break down cyclic di-GMP

Phospholipids Molecular structures that make up the membranes with a hydrophilic head and hydrophobic tails

Phosphorelay An extended form of a two-component regulatory system that has four proteins along which are passed a phosphate before the final response regulator binds to DNA

Phosphorylation The addition of a phosphate group $(PO_3)-$ to a protein

Photophores Specialized organs that host bacterial populations for the purposes of harnessing their light production

Physical map Determination of the base pair distances between restriction digest recognition sites in a chromosome

Pili Long thin bacterial surface structures made of protein and present in several copies per cell

Plasmid Small circular DNA that is separate from the chromosome(s), often containing non-essential traits

Plasmid copy number The number of plasmids present within a single bacterial cell

PMF *See* proton motive force

Point mutation A single base change in a DNA sequence

Polishing Improving the accuracy of sequencing data by comparing the final data to the original base calling data

Polyadenylation The addition of adenines to the 3′ end of mRNA

Polycistronic units mRNA with two or more genes

Polylinker *See* multiple cloning site

Polymerase chain reaction A molecular technique that copies specific fragments of DNA in the laboratory

Positive strand The strand of DNA encoding a protein

Post-segregational killing The death of a bacterial cell as a consequence of loss of a plasmid that encoded a toxin–antitoxin system; the toxin persists in the cell, but the antitoxin does not and the gene that encoded it was lost with the plasmid

Post-translational modification Changes made to a peptide chain after it is generated by translation of the mRNA at the ribosome

PPi *See* pyrophosphate

Pribnox box *See* −10 promoter sequence

Primary amino acid structure The linear sequence of the amino acids in a protein

Primary amino acid structure An unfolded string of amino acids joined together through the process of translation

Primase A replication enzyme that forms a short RNA primer necessary for DNA polymerase to be able to replicate the DNA

Processivity The property of a particular DNA polymerase to be able to continue to extend the DNA, generating the second strand of the helix against the template strand

Profile sequence In PSI-BLAST, the starting set of protein sequences that are closely related to each other, these are used as the basis for searches

Promoter occlusion When the transcription from one promoter negates or significantly reduces the transcription from a secondary promoter oriented in the same direction

Promoter region A location within DNA that is recognized by RNA polymerase for the initiation of transcription

Prophage The genome, in whole or part, of bacteriophage that has been integrated into the bacterial chromosome

Protein A polypeptide chain of amino acids

Protein fold The localized three-dimensional structures formed by amino acid sequences

Protein threading Generation of a three-dimensional model of a protein based on experimental evidence of the structure of proteins with similar fold structures; useful when there are no homologous proteins with identified structures

Proteinogenic amino acid An amino acid that can be found in proteins

Proteolysis Degradation of a protein

Proteome All of the proteins within a cell

Proteome All of the proteins present within a bacterial cell

Proteomics The study of the proteins present within a bacterial cell

Proton motive force The accumulation of protons on the surface of bacteria that forms an electrochemical gradient that can be used as an energy source for transport across the membranes

Pseudogene A genetic feature that has some, but not all characteristics of a coding sequence, suggesting that it may have been a gene that has degraded over time

Pseudopilus A protein component of a type 2 secretion system that pushes folded proteins out of the bacterial cell through the outer membrane complex

Pseudouridine A modified nucleoside present in some tRNAs

P-site *See* peptidyl site

PSK *See* post-segregational killing

Psychrophilic Bacteria that are capable of growth and cell division at extremely low temperatures

Purine The nitrogenous bases adenine and guanine, characterized by a pyrimidine ring fused to an imidazole ring

Pyrimidine The nitrogenous bases cytosine, thymine, and uracil, characterized by a pyrimidine ring

Pyrophosphate Two phosphates released from a nucleotide triphosphate during the incorporation of a base into DNA or RNA by the corresponding polymerase

Quaternary amino acid structure A complex protein structure formed from two or more folded tertiary protein structures

Query cover Short for query coverage, this BLAST statistic addresses how much of the query along its length has been found in the database

Query sequence The DNA, RNA, or protein data used to interrogate a database; generally the sequence the researcher is interested in investigating

Quorum sensing The production and sensing of a chemical signal that allows the bacteria to determine the size of its population

R When discussing amino acids, the R stands for any side chain

Random mutagenesis The generation of mutants that not directed, but are randomly generated due to the methods used

Raw sequencing data Information collected from a sequencing technology machine prior to removal of poor quality data

RBS *See* ribosome binding site

Read length This is a measure of how many nucleotides have been identified in each individual sequencing reaction

Reading frame Each set of three bases that makes up a codon in DNA; being read forward or reverse there are therefore six reading frames starting from sequential bases on the forward or reverse strand

Receiver domain The section of a response regulator protein in a two-component regulatory system that is phosphorylated by the histidine kinase

Reciprocal best hits Two sequences that are the best hits for one another in homology

Recombinant DNA Two DNA molecules from different sources that are combined into one DNA molecule in the laboratory

Recombinant protein A protein that is generated from recombinant DNA, such that the protein is made by a different species than the origin of the DNA encoding the protein

Recombination The exchange of DNA

Reductive genome evolution A process of eliminating genetic content that is not needed by the bacterial cell

Regions of difference Regions of the DNA that are different between two or more sequences that are otherwise similar

Regions of difference Sections of a bacterial genome sequence that are different from what is found in other related bacterial genome sequences

Regulated expression Expression of a transcript of protein that is controlled so that there can be different levels of expression within the cell

Regulator A protein that can alter transcription of a gene

Regulatory network Connection between multiple genes in their expression due to common regulatory proteins and/or ncRNAs

Regulon The set of genetic features controlled by a common regulator

Release factor A protein that enters the A-site at the termination of translation and triggers disassociation of the translation machinery

Replication In genetics, the process of copying DNA, generally making a copy of a genome or chromosome

Replication fork During the process of replication, the two DNA strands are split apart; this is the split that moves along the chromosome as replication progresses

Repressor A protein that can stop or reduce transcription of a gene

Repressor protein A protein that can stop or reduce the transcription of a gene or genes, often through binding to DNA

Resequencing Conducting sequencing on a sample or bacterial species that has already been sequenced, often to determine changes that may have occurred over time

Response regulator A protein in a two-component regulatory system that is able to bind to DNA when it is phosphorylated

Restriction endonucleases *See* restriction enzymes

Restriction enzymes Proteins that are able to cleave DNA at a specific site based on recognition of a specific sequence in the DNA

Restriction fragments Pieces of DNA generated by enzymatic digestion with a restriction enzyme

Restriction map *See* physical map

Restriction modification system A restriction enzyme with a specific recognition sequence and a methylase that is able to methylate that recognition sequence to protect it from the restriction enzyme

Reverse vaccinology The design and development of a vaccine based on genome sequencing data and predicted antigens

Rho-dependent transcriptional terminator When the signal to stop transcription relies on the binding and action of a separate protein called Rho

Ribonucleic acid (aka RNA) genetic material used by bacteria to express their traits, comprised of ribose sugars, phosphates, and the nitrogenous bases adenine, cytosine, guanine, and uracil

Ribonucleotide triphosphate A ribose sugar, three phosphates, and a nitrogenous base

Ribose A sugar containing five carbons, five oxygens, and 10 hydrogens that forms the backbone of RNA, together with phosphate

Ribosomal RNA A type of ribonucleic acid that forms a complex with ribosomal proteins to form a ribosome, wherein translation of mRNA occurs

Ribosome A complex of proteins and ribosomal RNA essential for translating mRNA to protein

Ribosome binding site A location on mRNA where ribosomes bind to initiate translation

Ribosome recycling In genetics, the final phase of translation where the released ribosome is able to be involved in another translation initiation

Riboswitch Encoded regulatory factor for transcription or translation that alters expression by virtue of its sequence

Riboswitch A portion of the mRNA transcript that participates in the regulation of the expression of the mRNA containing it

Ribozyme An RNA with enzymatic activity

RNA *See* ribonucleic acid

RNA polymerase An enzyme that is able to form a new strand of RNA based on a single strand of DNA as the template

RNA thermometers mRNA sequences that alter in their secondary structure in response to temperature changes, which affects expression of the encoded protein(s)

RNA World Hypothesis A scientific idea that RNA was used as the genetic material on Earth before there was DNA

RNase Enzymes that degrade RNA

RND family A class of efflux pump made of three proteins, an inner membrane protein, membrane fusion protein, and outer membrane protein

Rough A description that can be applied to Gram-negative bacteria with full-length LPS O-antigen structures

rRNA *See* ribosomal RNA

Salmonella-containing vacuole A specialized host cell organelle created through the action of Salmonella Type 3 Secretion System effector proteins from the host cell vacuole specifically taking up a Salmonella cell

Saprophyte An organism that lives on decaying or dead organic matter

Scaffold Draft or incomplete sequence data that may contain unidentified nucleotides or errors that can be used to build more accurate sequence information by virtue of its relative locational information

Screening In molecular genetics, this is the process of identifying bacterial cells with a particular phenotype by testing each clone for a particular trait

SCV *See* Salmonella-containing vacuole

Sec pathway *See* Sec translocase

Sec translocase A protein system for transporting unfolded peptide chains across the cytoplasmic membrane

Secondary amino acid structure Basic structures formed by primary amino acid sequences, including the α helix and β sheet

Secondary amino acid structure A string of amino acids that have a simple structure that is formed by hydrogen bonding between amino acids in the chain

Secretion ATPase A protein component of a type 2 secretion system that provides energy from ATP in the cytoplasm

Secretion system Cellular machinery that exports proteins and other substrates out of the bacterial cell

Selection In molecular genetics, this is the process of identifying bacterial cells with a particular phenotype by generating media on which only the desired clones will grow

Selenocysteine Considered the 21st amino acid, it is encoded by a stop codon and other sequence features

Sense strand *See* positive strand

Sensor kinase A membrane protein in a two-component regulatory system that autophosphorylates in response to stimuli

Sequence The order of nitrogenous bases in a DNA or RNA

Sequencing by synthesis Determination of the sequence of a nucleic acid using DNA polymerase to synthesize the second strand and monitor the incorporation of nucleotides

Sequencing platform A phrase that refers to the sequencing equipment, technology, and reagents collectively

Sessile cells Cells that are not motile, but rather are fixed in one place, generally specifically referring to one of two stages of the life cycle of *Caulobacter*

β sheet *See* beta sheet

Shine–Dalgarno sequence A conserved sequence generally found 5′ of genes, which is recognized by the 16S rRNA within the ribosome

Shotgun sequencing Genome sequencing methodology that starts with random fragmentation of the genome before sequencing, with the goal of assembling the random sequence data using overlapping sequence data

Siderophore A molecule able to bind iron and scavenge it for the bacteria from the host environment

Sigma subunit One of six proteins that come together to form RNA polymerase; this subunit dictates promoter specificity

Signal recognition particle A series of amino acids at the start of a peptide chain that are recognized for transport across the membrane

Signal sequence A specific amino acid sequence in a peptide chain that is recognized for translocation of the protein across the membrane

Signature tagged mutagenesis A modification of transposon mutagenesis where a transposon library is made with each transposon carrying a unique signature tag that can be used for identification

Similarity (amino acid) A measure, usually expressed in percentage, of how much two proteins carry amino acids with attributes that match

Single-molecule sequencing The ability to generate sequencing data from a single molecule of DNA without amplification

Single-nucleotide polymorphism A change in one base of a DNA sequence

Small RNA A short piece of RNA that can serve a regulatory function within the cell

SmartBLAST A specialized BLAST that outputs the five best protein matches from a specially curated database of reference sequences, *see* landmark database

Smooth A description that can be applied to Gram-negative bacteria with short LPS O-antigen structures

SMR family A class of efflux pump involved in multidrug resistance that is made of several copies of a single protein

S-nitrosylated The addition of a thiol group (−SNO) to a protein

SNP *See* single-nucleotide polymorphism

Sonication Use of sound waves to disrupt a sample, including breaking open biological membranes to release proteins and DNA

Specialized transduction Due to imprecise excision of the prophage from the bacterial chromosome, some bacterial DNA may be packaged into the bacteriophage head and therefore carried to new bacterial cells

SRP *See* signal recognition particle

Star activity Nonspecific, off-target cleavage of DNA by restriction enzymes due to variations in time and concentration

Start point of transcription The location on DNA where the generation of RNA based on the DNA sequence is started

Stem loop A secondary structure formed by DNA or RNA whereby complementary bases on the same strand

hydrogen bond to one another, making the stem, and bending at non-complementary bases, making a loop

Stereoisomers Compounds containing the same elements, which are mirror images of each other structurally

Steric hindrance When the structures of two structures interfere with one another and therefore limit the form of the overall structure

Sticky ends Ends of a DNA double helix generated by a restriction enzyme, where the DNA cut is staggered so that there are overhanging nucleotides on either the 5′ or 3′ ends

Stop codon One of three specific three-letter codes that signal the termination of translation

Strain A representative of a bacterial species that may slightly vary genetically from other members of the same species

Stringent response A regulated stress response of bacteria due to factors such as nutrient limitation, amino acid starvation, heat shock, iron depletion, or other stresses

σ subunit *See* sigma subunit

Succinylation The addition of a succinyl group ($-CO-CH_2-CH_2-CO_2H$) to a protein

Supercoil The twisting of a DNA double helix upon itself, a process that compacts and condenses the DNA

Surveillance Tracking and monitoring of infections within a clinical setting, such as a hospital

Swarmer cells Cells that are motile, possessing flagella to enable motility, generally specifically referring to one of two stages of the life cycle of *Caulobacter*

Symbiosis A long-term interaction between two different bacterial species

Synonymous A change in a DNA sequence that results in the same amino acid when translated as it would have done before the change

Syntenic Describing sequences with synteny

Synteny Similarity between two or more sequences in the organization of the region, including genes, similar sequences, and other genetic features

Synteny A term used to describe regions that are similar in the order of the features they contain, regardless of sequence similarity

T1SS *See* type 1 secretion system

T2SS *See* type 2 secretion system

T3SS *See* type 3 secretion system

T4SS *See* type 4 secretion system

T5SS *See* type 5 secretion system

T6SS *See* type 6 secretion system

T7SS *See* type 7 secretion system

Target sequence The sequence from a database that is hit upon by a search, such as BLAST, as being similar to the query sequence

Tat pathway A system for the transport of folded proteins across the cytoplasmic membrane

TBLASTN *See also* BLAST, a specialized form of BLAST for comparing an amino acid sequence with a nucleotide sequence database by first translating the nucleotide sequence database into the six frame protein translation amino acid sequences for each database entry

TBLASTX *See also BLAST*, a specialized form of BLAST for comparing a nucleotide sequence with a nucleotide sequence database, but which first generates six frame protein translation amino acid sequences for both the query sequence and each database entry and then compares these protein translations

Temperature-sensitive mutant A bacterial cell with a mutation that only manifests when there is a change in temperature

Termination (transcription) In genetics, the end of the process of transcription, where transcription enzymes disassociate in response to a signal to stop transcription

Termination (translation) In genetics, the end of the process of translation, where the assembled components disassociate and translation stops

Termination codon *See* stop codon

Termination of replication The region of the chromosome where a round of replication ends

Terminus In bacterial genetics, the location on the chromosome where replication ends

Tertiary amino acid structure Complex protein structures formed when secondary structures come together through hydrogen bonding, side chain interactions, ionic interactions, and disulfide bonds

Tertiary amino acid structure A string of amino acids that is folded into a complex structure

Third base wobble The recognition that the third base of a codon can change but the encoded amino acid can stay the same

Thymine A nitrogenous base made of five carbons, two nitrogens, two oxygens, and six hydrogens, found in DNA and RNA

TLD *See* tRNA-like domain

T-loop The loop from the third of three-stem loop structures in tRNA, starting from the 5′ end

tmRNA *See* transfer messenger RNA

Tn-seq The sequencing of a transposon library to reveal the regions of the genome required for various growth conditions

Tolerance Due to the physiological state of a population of bacteria, often in a biofilm, a proportion of the bacterial cells are able to survive antimicrobial killing

Tolerant *See* tolerance

Tot score Short for total score, this takes account of a BLAST query having similarity to a subject in multiple regions

Trans Referring to chemical structures those being opposite one another; referring to genetics those being distant

Transcript The product of transcription, RNA made by RNA polymerase based in a DNA template

Transcription In genetics, the process whereby DNA is copied into RNA

Transcription interference When the transcription from one promoter influences the transcription from another promoter

Transcriptional regulation The process of altering transcription to increase or decrease the levels of RNA produced

Transcriptome All of the RNA transcripts within a cell

Transduction The transfer of DNA from one bacterial cell to another via a bacteriophage

Transfer messenger RNA A specialized RNA within the bacterial cell that has activity like both a tRNA and mRNA and which releases the translation machinery stalled in non-stop decay and tags the peptide for degradation

Transfer RNA A type of ribonucleic acid that forms a specific structure capable of binding to an amino acid and of base pairing with the corresponding codon on a mRNA

Transformation The uptake of DNA from the environment by a bacterial cell, which may then incorporate the DNA into its own

Translation In genetics, the process whereby an mRNA sequence is used to assemble an amino acid sequence and form a protein

Translation initiation region The region on an mRNA that is recognized by a ribosome to begin translation

Translocation The movement of DNA from one location to another

Translocon A component of a type 3 secretion system that is at the tip of the needle and regulates the release of effectors

Transposon mutagenesis The use of transposons to disrupt the sequence of a bacterial chromosome, generating mutants

Transposons Features within bacterial DNA that are capable of self-mobilization from one part of the DNA to another and which carry additional genes

Transposon-sequencing *See* Tn-Seq

Trans-translation The process of completing translation where there is no termination codon on the mRNA, using tmRNA that tags the erroneous protein for degradation and terminates translation

tRNA *See* transfer RNA

tRNA-like domain Part of a tmRNA that is like tRNA and able to carry an amino acid to the ribosome

Tryptophan One of the amino acids within the bacterial cell that is incorporated into proteins

T-stem The stem from the third of three-stem loop structures in tRNA, starting from the 5′ end

twin-Arg pathway *See* Tat pathway

Twitching motility Movement of bacterial cells along surfaces via the retraction and extension of pili

Two-component regulatory system A set of two proteins that regulate transcription in response to stimulus

Two partner secretion A type 5 secretion system where two proteins work to make an outer membrane channel and export the partner protein

Two-component regulatory system A system for transcriptional regulation that uses a protein to sense stimuli and a second protein to regulate transcription based in input from the sensor

Type 1 secretion system A three-protein system with a ABC transporter, membrane fusion protein, and outer membrane protein that is capable of exporting substrates from Gram-negative bacteria

Type 2 secretion system A four-protein system with an inner membrane platform, secretion ATPase, pseudopilus, and outer membrane complex that is capable of exporting substrates from Gram-negative bacteria

Type 3 secretion system A three-part system with a base complex, needle, and translocon made of several proteins that is capable of exporting substrates from Gram-negative bacteria

Type 4 secretion system A multi-protein system with homology to bacterial pili that is capable of exporting substrates from Gram-negative bacteria

Type 5 secretion system Proteins that are capable of transporting themselves out of the Gram-negative bacterial cell

Type 6 secretion system A multi-protein system with homology to bacteriophage that is capable of exporting substrates from Gram-negative bacteria

Type 7 secretion system A system that aids in exporting substrates through the thick peptidoglycan layer of Gram-positive bacteria

Untranslated region The segment of mRNA 5′ of the initiation codon that is not translated, yet is part of the mRNA

Untranslated region A portion of the mRNA transcript that is not translated by the ribosome into protein

UP element A conserved sequence 5′ of a transcriptional start site that enhances the binding of RNA polymerase

Upstream Before a gene or other genetic feature, 5′ of the feature

Uracil A nitrogenous base made of four carbons, two nitrogens, two oxygens, and four hydrogens, found in RNA

UTR *See* untranslated region

UTR *See* untranslated region

Vacuole A small membrane-bound space within a eukaryotic cell

Variable loop A loop that is present in some tRNAs, located between the anticodon stem and the T-stem

Virion The protein coat of a bacteriophage particle

Xeno nucleic acid An artificially created nucleic acid with a structure not found in nature, generally with sugars other than ribose or deoxyribose

XNA *See* xeno nucleic acid

Zoonotic disease Infectious diseases that primarily infect animals, but can be transferred to humans as well

Zwitterion A molecule that has a negative charge at one end, a positive charge at the other end, and a net zero charge

Glossary of Bacterial Species

Bacteria	What they do	Gram stain
Acinetobacter baumannii	An opportunistic human pathogen, associated with hospital-acquired infections. Of concern in immunocompromised individuals, particularly due to multi-drug resistance.	Gram-negative
Actinobacillus pleuropneumoniae	A pathogen in pigs causing a frequently lethal necrotizing hemorrhagic pneumonia.	Gram-negative
Agrobacterium tumefaciens	A plant pathogen that is capable of infecting a wide range of plants, causing tumor-like growth referred to as crown-gall diseases.	Gram-negative
Anaeromyxobacter dehalogenans	A non-pathogenic bacterial species present in soil and important for understanding soil dynamics.	Gram-negative
Anaplasma marginale	A pathogen of cattle and other ruminant animals, which is transmitted by ticks, causing progressive hemolytic anemia and death.	Gram-negative
Bacillus anthracis	A pathogen that causes three distinct types of human infection (cutaneous, inhalation, and gastrointestinal) and is capable of forming spores.	Gram-positive
Bacillus subtilis	A non-pathogen, spore-forming bacteria that has been isolated from a variety of environments, including the guts of animals, oceans, and soil.	Gram-positive
Bacteroides fragilis	Bacteria that are normally resident in the human gut as commensals, but can cause infection following surgery, trauma, or disease.	Gram-negative
Bacteroides ovatus	Bacteria that are normally resident in the human gut as commensals, but appear to be involved in antibody responses in inflammatory bowel disease.	Gram-negative
Bordetella pertussis	A pathogen of humans that causes whooping cough, which can have fatal complications, including bronchopneumonia and acute encephalopathy.	Gram-negative
Borrelia burgdorferi	Transmitted by ticks, these bacteria cause Lyme disease in humans as a pathogen, but is not pathogenic in the small mammal reservoirs for these bacteria.	Gram-negative
Buchnera aphidicola Buchnera aphidicola	A bacterial endosymbiont of aphids, these Gram-negative bacteria do not have LPS or LOS.	Gram-negative
Burkholderia cepacia	Bacteria that frequently colonize fluids used in hospital wards, they pose health threats to hematology and oncology patients.	Gram-negative
Burkholderia dolosa	Bacteria that can be fatal in patients with cystic fibrosis.	Gram-negative
Burkholderia multivorans	Bacteria that can cause chronic disease and infection, particularly in patients with cystic fibrosis or other immunocompromised states.	Gram-negative
Burkholderia pseudomallei	The bacterial cause of the disease melioidosis, which can be potentially fatal.	Gram-negative
Campylobacter coli	A pathogenic bacterial species causing food poisoning in humans, including diarrhea.	Gram-negative
Campylobacter jejuni	One of the major causes of bacterial food poisoning from poultry, which not only causes significant disease, but can be fatal.	Gram-negative
Candidatus Zinderia insecticola	A symbiotic bacteria that lives within the spittlebug *Clastoptera arizonana*, it has a very small genome of 208 kb with 13.5% G+C.	Gram-negative
Candidatus Nasuia deltocephalinicola	A symbiotic bacteria that lives within the leafhopper insect *Macrosteles quadripunctulatus* and has a very small genome of 112 kb and 16.6% G+C.	Gram-negative
Caulobacter	A genus of bacteria that live in aquatic environments and are model organisms for the study of the cell cycle and synchronized cultures.	Gram-negative
Chlamydia pneumoniae	An obligate intracellular pathogen that is a common cause of bacterial pneumonia.	Gram-negative
Chlamydia trachomatis	An obligate intracellular pathogen that is the cause of the sexually transmitted disease chlamydia, amongst other diseases.	Gram-negative
Clavibacter michiganensis	A bacterial plant pathogen of economic importance due to the impact of disease on food crops such as potatoes and tomatoes.	Gram-positive

Bacteria	What they do	Gram stain
Clostridioides difficile	A pathogen of humans that is a major cause of hospital-acquired diarrhea; capable of forming spores. Formerly *Clostridium difficile*.	Gram-positive
Clostridium perfringens	A common cause of food poisoning, which causes abdominal cramps and diarrhea; capable of forming spores.	Gram-positive
Clostridium spp.	A genus of bacteria that are spore forming and typically anaerobic, some of which can opportunistically cause human disease. *Clostridioides difficile* was previously classified within this genus.	Gram-positive
Clostridium stercorarium	A bacterial species of interest for exploitation for biotechnology due to its ability to degrade polysaccharides in plant biomass, resulting in ethanol and acetate.	Gram-positive
Corynebacterium diphtheriae	A pathogen of humans that causes diphtheria, expressing diphtheria toxin.	Gram-positive
Coxiella burnetti	A pathogen that causes Q fever in humans and coxiellosis in animals.	Gram-negative
Dechloromonas aromatic	Bacteria of biotechnology interest because of their ability to oxidize benzene, a difficult pollutant to address.	Gram-negative
Enterobacter spp.	A genus of bacteria predominantly found in the intestines of animals that are able to thrive in aerobic and anaerobic environments; most have flagella.	Gram-negative
Enterobacteriaceae	A family of bacteria that includes species capable of causing disease in humans and animals, found in intestines as well as other environments.	Gram-negative
Enterococcus faecalis	A bacterial pathogen that can cause hospital-acquired urinary tract, wound, and other infections.	Gram-positive
Enterococcus faecium	A bacterial species that is a leading cause of nosocomial infections, especially in critically ill or immunocompromised patients.	Gram-positive
Escherichia coli	A diverse genus of bacteria studied as a model in bacterial genetics and genomics, used in biotechnology, which is sometimes pathogenic.	Gram-negative
Francisella novicida	A non-pathogenic species of interest due to its genetic similarity to the pathogen *Francisella tularensis*.	Gram-negative
Geobacter sulfurreducens	A bacterial species of interest for exploitation for biotechnology in generating fuel cells, bioremediation, and biofuel production.	Gram-negative
Haemophilus ducreyi	A bacterial pathogen that causes the sexually transmitted infection chancroid.	Gram-negative
Haemophilus influenzae	A species of bacteria that can cause a variety of human diseases, including meningitis, septicemia, and pneumonia.	Gram-negative
Haemophilus parainfluenzae	An opportunistic human pathogen that can cause some similar, yet less severe infections as *Haemophilus influenzae*.	Gram-negative
Helicobacter canadensis	A zoonotic bacterial species that is an emerging pathogen in humans, causing diarrhea.	Gram-negative
Helicobacter pylori	A bacterial pathogen of humans that is able to colonize and survive for decades in the stomach and upper digestive tract; cause of ulcers and other diseases.	Gram-negative
Klebsiella pneumoniae	A pathogen of humans that can harmlessly colonize the gut, but causes hospital-acquired and community-acquired pneumonia.	Gram-negative
Lactobacillus	A genus of bacteria that produce lactic acid in their metabolism, they are found associated with animals, plant material, and fermented or spoiled foods.	Gram-positive
Lactobacillus plantarum	A bacterial species of interest to the food industry as a probiotic, it survives passage through the intestinal tract and produces lactic acid.	Gram-positive
Legionella pneumophila	A pathogen of humans that causes legionellosis, an acute pneumonia.	Gram-negative
Leptospira spp.	A genus of bacteria that cause leptospirosis, zoonotic disease. Can cause a wide range of symptoms and be underdiagnosed.	Gram-negative
Listeria monocytogenes	A pathogen associated with deli meats and cheeses, of concern to pregnant women, people over 65, and immunocompromised individuals.	Gram-positive
Mannheimia haemolytica	An opportunistic pathogen causing pneumonia in cattle and calves.	Gram-negative
Moraxella bovis	A pathogen causing keratoconjunctivitis (eye infection) in cattle with symptoms that range from mild to severe.	Gram-negative

Bacteria	What they do	Gram stain
Moraxella catarrhalis	A pathogen of humans causing otitis media (middle ear infection) and exacerbations of COPD (chronic obstructive pulmonary disease).	Gram-negative
Moraxella lancunata	A pathogen of humans that can cause eye infections and in rare cases some serious infections, usually associated with other stress or immunocompromise.	Gram-negative
Mycobacterium bovis	A pathogen found in cattle and related animals that can also cause tuberculosis in humans.	Acid fast[a]
Mycobacterium leprae	A pathogen that damages peripheral nerves and can cause severe deformities.	Acid fast[a]
Mycobacterium tuberculosis	A pathogen that causes tuberculosis as well as other human diseases, which can result in death.	Acid fast[a]
Mycoplasma fermentans	An opportunistic pathogen of immunocompromised humans, often seen as a co-infection associated with other diseases.	Gram-negative
Mycoplasma genitalium	A pathogen causing sexually transmitted infections in humans, which may have symptoms similar to gonorrhea or chlamydia, or no symptoms.	Gram-negative
Mycoplasma hominis	A pathogen associated with other bacteria in pelvic inflammatory disease and bacterial vaginosis, and also causing infections related to pregnancy.	Gram-negative
Mycoplasma gallicepticum	A pathogen causing respiratory infections in chickens and turkeys, as well as other birds.	Gram-negative
Mycoplasma pneumoniae	A pathogen that can cause mild pneumonia, sometimes referred to as "walking pneumonia," as well as more severe forms.	Gram-negative
Neisseria gonorrhoeae	A pathogen in humans that causes the sexually transmitted infection gonorrhea and infects other mucosal sites.	Gram-negative
Neisseria lactamica	A non-pathogenic commensal that colonizes the human nasopharynx; closely related to pathogens *N. gonorrhoeae* and *N. meningitidis*. A non-pathogenic commensal that colonizes the human nasopharynx; closely related to pathogens *N. gonorrhoeae* and *N. meningitidis*.	Gram-negative
Neisseria meningitidis	A pathogen in humans that colonizes the nasopharynx and can cause life-threatening invasive meningococcal meningitis and septicemia.	Gram-negative
Paracoccus denitrificans	A bacterial species from soil that can grown aerobically or anaerobically on organic or inorganic energy sources. It may be the ancestor of eukaryotic mitochondria.	Gram-negative
Pseudoalteromonas haloplanktis	A bacterial species isolated from a sample of Antarctic sea water, which is able to live in cold temperatures.	Gram-negative
Pseudomonas aeruginosa	An opportunistic human pathogen, particularly of concern in immunocompromised individuals, it is commonly found in a variety of environments.	Gram-negative
Pseudomonas extremaustralis	A bacterial species from the Antarctic that is able to grow at low temperatures and is resistant to stress and high levels of polyhydroxybutyrate.	Gram-negative
Rhodobacter sphaeroides	A rod shaped, metabolically diverse bacterial species that has potential in the development of biofuels, bioplastics, and bioremediation.	Gram-negative
Rickettsia prowazekii	Intracellular bacteria that cause the human disease typhus, transmitted by the human body louse.	Gram-negative
Rickettsia rickettsii	Intracellular bacteria that cause the human disease Rocky Mountain Spotted Fever, transmitted by ticks.	Gram-negative
Rickettsia spp.	A genus of intracellular bacteria that are responsible for spotted and typhus fevers.	Gram-negative
Salmonella enterica serovar *Enteritidis*	A pathogen causing illness following consumption of infected food, particularly poultry and eggs.	Gram-negative
Salmonella enterica typhimurium	A pathogen infecting humans and animals following ingestion of contaminated food or water.	Gram-negative
Serratia marcescens	An opportunistic pathogen that can cause a range of infections in humans and often produces a red pigment.	Gram-negative
Shewanella oneidensis	A bacterial species of interest in bioremediation due to its ability to reduce metals, including those in nuclear waste, causing uranium to precipitate out of water.	Gram-negative
Shigella spp.	A genus including bacterial pathogens that cause shigellosis, severe diarrhea that may contain blood or mucus.	Gram-negative

Bacteria	What they do	Gram stain
Sodalis glossinidius	An endosymbiont of the tsetse fly that has a mutualistic relationship with its host organism.	Gram-negative
Sorangium cellulosum	A species of bacteria with large genomes that produce secondary metabolites of interest in drug discovery and development.	Gram-negative
Staphylococcus aureus	A pathogen of humans that can live harmlessly on the skin, yet cause serious, life-threatening infections within the body.	Gram-positive
Staphylococcus epidermidis	An opportunistic pathogen that is commonly associated often with catheters and other indwelling devices.	Gram-positive
Streptococcus agalactiae	A pathogen, also known as GBS or group B Strep, that causes serious infections in newborn infants.	Gram-positive
Streptococcus pneumoniae	A pathogen of humans that causes pneumonia when the bacteria enter the lungs via inhalation.	Gram-positive
Streptococcus pyogenes	A pathogen, also known as GAS or group A Strep, that causes a range of infections in humans of varying severity.	Gram-positive
Streptococcus uberis	A pathogen of dairy cows that causes mastitis (infection of the udders).	Gram-positive
Streptomyces coelicolor	A spore-forming, soil-dwelling bacterial species that produces a variety of antibiotics.	Gram-positive
Thermus aquaticus	A bacterial species isolated from hot springs in Yellowstone National Park, capable of growth at high temperature, of significance when its thermostable DNA polymerase was used for PCR.	Gram-negative
Thiomargarita namibiensis	A bacterial species that was isolated off the Namibian coast and found to be much larger than most.	Gram-negative
Tropheryma whipplei	A pathogen of humans that causes Whipple's diseases as well as other infections.	Gram-positive
Ureaplasma	A genus of bacteria able to hydrolyze urea and can cause disease in humans; do not have a cell wall.	Gram-negative[b]
Vibrio cholerae	A pathogen of humans that lives in the aquatic environment, but causes diarrhea and vomiting upon ingestion by humans.	Gram-negative
Aliivibrio fischeri	An aquatic bacteria capable of bioluminscence that can be found living symbiotically with marine animals such as the bobtail squid. Formerly *Vibrio fischeri*.	Gram-negative
Vibrio harveyi	An aquatic bacteria capable of bioluminscence that is a pathogen of marine organisms.	Gram-negative
Vibrio parahaemolyticus	A pathogen of humans causing diseases associated particularly with consumption of raw seafood.	Gram-negative
Wolbachia	A genus of bacteria that infect insects, having complex interactions with their hosts.	Gram-negative
Yersinia pestis	A pathogen of humans and animals that also infects fleas, which transmit the bacteria; cause of plague.	Gram-negative
Yersinia pseudotuberculosis	A zoonotic bacterial species that is a human pathogen causing disease following ingestion of food or water contaminated with animal feces.	Gram-negative

[a] The cell surface of this species means that it does not absorb Gram stain and cannot be classified by Gram staining.
[b] Stain as Gram-negative due to the absence of a cell wall.

Index

Taylor & Francis Group
an **informa** business

Taylor & Francis eBooks

www.taylorfrancis.com

A single destination for eBooks from Taylor & Francis
with increased functionality and an improved user
experience to meet the needs of our customers.

90,000+ eBooks of award-winning academic content in
Humanities, Social Science, Science, Technology, Engineering,
and Medical written by a global network of editors and authors.

TAYLOR & FRANCIS EBOOKS OFFERS:

A streamlined
experience for
our library
customers

A single point
of discovery
for all of our
eBook content

Improved
search and
discovery of
content at both
book and
chapter level

REQUEST A FREE TRIAL
support@taylorfrancis.com

Routledge
Taylor & Francis Group

CRC Press
Taylor & Francis Group